RACE, RELIGION, AND A CURRICULUM
OF REPARATION

RACE, RELIGION, AND A CURRICULUM OF REPARATION

Teacher Education for a Multicultural Society

William F. Pinar

Softcover reprint of the hardcover 1st edition 2006 978-1-4039-7072-5

First published in 2006 by
PALGRAVE MACMILLAN™
175 Fifth Avenue, New York, N.Y. 10010 and
Houndmills, Basingstoke, Hampshire, England RG21 6XS
Companies and representatives throughout the world.

PALGRAVE MACMILLAN is the global academic imprint of the Palgrave Macmillan division of St. Martin's Press, LLC and of Palgrave Macmillan Ltd. Macmillan® is a registered trademark in the United States, United Kingdom and other countries. Palgrave is a registered trademark in the European Union and other countries.

ISBN 978-1-349-53242-1   ISBN 978-1-4039-8473-9 (eBook)
DOI 10.1057/9781403984739

Library of Congress Cataloging-in-Publication Data

Pinar, William.
    Race, religion, and a curriculum of reparation : teacher education for a multicultural society / William F. Pinar.
        p. cm.
    Includes bibliographical references and index.

    1. Bible. O.T. Gen. IX, 20–25—Criticism, interpretation, etc. 2. Blacks—Biblical teaching. 3. Schreber, Daniel Paul, 1842–1911. 4. Sex in the Bible. I. Title.

BS1235.52P56 2006
305.800973—dc22   2005054218

A catalogue record for this book is available from the British Library.

Design by Newgen Imaging Systems (P) Ltd., Chennai, India.

First edition: April 2006

10 9 8 7 6 5 4 3 2 1

Transferred to Digital Printing 2011

Contents

PREFACE

[W]hite ethnicity constitutes an "unknown" in contemporary cultural theory—a dark continent that has not yet been explored. Kobena Mercer (1994, 217)

[I] see resistance as a way of politically activating counter-memories, that is to say sites of non-identification with or non-belonging to the phallogocentric regime. Rosi Braidotti (1994b, 201)

[T]he study of racism is dirty business. It unveils things about ourselves that we may prefer not to know. Lewis R. Gordon (1995, ix)

[O]ur deepest cultural assumptions are biblical. Regina M. Schwartz (1997, x)

This is a textbook for teachers, a synoptic text summarizing and juxtaposing research that enables teachers to complicate the curricular conversation in which they and their students are engaged. The book is a form of curriculum research, less concerned with (although hardly disinterested in) pedagogy than with academic knowledge. As in intellectual history, this form of curriculum research appreciates that "understanding always therefore entails what might be called . . . proleptic paraphrase or anticipatory synopsis" (Jay 1988a, 59). That is, by juxtaposing fragments from various disciplinary traditions, I support students' study of race from vantage points anticipated, perhaps, by no one discipline.

Situated between disciplines, then, such synoptic curriculum research proceeds without the usual discursive sanctions provided by disciplinarity. With Kaja Silverman (2000, 27),

> I am thus obliged to acknowledge what I might otherwise disavow: my discourse is as groundless as desire itself. What I have to offer is only what can be seen from the finite and singular perspective that this vantage point opens up. Others will be able to apprehend what I cannot apprehend: the many perspectives which mine works to close-off.

Within the structure of this textbook, I seek to stage and thereby provoke complicated conversation.

I continue to work on understanding race, in particular, whiteness. that I seek its origins. The textbook for teachers that I prepared on the gender of racial politics and violence in America concluded with a reference to the genesis of "race" in the West. Slaveholders and the segregationists who followed them justified their practices by references to the Bible, a practice not uncommon in the American South today, although gender, not race, is the salient subject of biblical injunction today. The substitution is no accident.

For Confederates past and present, God himself ordained the "Great Chain of Being" at which "whites" are secured at the top. In the United States, this racial hierarchy is not only social and economic, it is gendered, indeed, sexualized. Bleached of race (conservatives insist that racism is past), the civic sphere is specularized and sexualized. Until the conflation of "race" and "sex" becomes unknotted, we educators cannot teach "tolerance." To untie that knot, I return to the primal scene of race, Genesis 9:24.

What happened that mythic night in Noah's tent? The main points are these: Noah (of flood fame) plants a vineyard, makes wine, gets drunk, and passes out, naked, in his tent. His son Ham—Noah has two other sons (Shem and Japheth)—goes into the tent and, later, leaves. After some time passes, Noah emerges: "And Noah awoke from his wine, and knew what his younger son had done unto him" (Genesis 9:24). Noah curses not Ham, but Ham's son Canaan: "a servant of servants shall he be unto his brethren" (Genesis 9:25). There is no explicit reference to "race" in these passages. Although Ham connotes "dark" in ancient Hebrew, it would seem that slaveholders and segregationists fabricated the association of race with Noah's rage.[1] Why?

To answer that question, we must return to the passage, to the provocation of the rage. What could Ham have done to prompt Noah to curse his grandson and his progeny into perpetual enslavement? Exegetes have proposed two main answers. The first and primary one is that Ham violated the ancient Israelite prohibition against looking at the body of the father. The secondary answer is that Ham violated his father sexually. These are intersecting speculations, I suggest, as each involves a symbolic "castration" of the father, an "unmanning" that the patriarch repudiates by cursing his son's son to servitude. In his defensive rejection of the son's desire (Noah, the second Adam, almost "replants" the Garden in his marriage tent or Chuppah), Noah curses Ham's progeny to servitude.

The "Noah Complex" is Regina M. Schwartz's (1997, 115) phrase, devised to depict the dynamics of the curse of Ham. Schwartz (1997, 115) points out that the son's incestuous desire for the father produces an "intolerable guilt" that is projected heavenward. God-the-Father (or his emissary, in this case Noah) punishes the son (or grandson, in this case) by turning him into a "reviled Other." Schwartz invokes the notion of "scarcity" to account for this curse, scarcity referring to a "shortage of parental blessings and love" (1997, 115). "Scarcity imposes hierarchy," she continues, it "imposes patriarchy" (1997, 115–116).

Scarcity is, I suggest, a misleading term to depict the genesis of hierarchy, of patriarchy, of "race." It is not "scarcity" but a surplus of desire that provokes the father to sacrifice his son for the sake of the social order. True, scarcity describes the emotional poverty of Yahweh, a demanding and unforgiving Father who commands his son to cut his penis in order to demonstrative his piety. "The son's desire for his father,"

Schwartz (1997, 114) understands,

> is expressed in efforts to become like the father (in his image), in yearnings to build heavenward, yearnings to become "as the gods," yearnings not only for the father's blessing but for the father's mantle, yearnings to enter the presence of God as Moses does and to be transformed into radical by his glory, and even yearnings to *be* God, as in the case of Christ.

Desire structures such yearning. "For masculine identity to be saved," literary scholar Scott S. Derrick (1997, 59) knows, "the nature of its constitution must be forgotten."

Servitude severs the self, now self-split and abjected, an "other." "In other words," David Marriott (2000, 12) writes (in a different but related context), "the violated body of the black man comes to be used as a defense against the anxiety, or hatred, that body appears to generate." The tracing of white racism to its origins in gender, specifically in the repudiation of father–son desire, functions to subvert its curse; it risks but not does not achieve an "evacuation" of the "significance" of race, as Robyn Wiegman (1995, 163) worries. Rather, it implies a shattering of white (especially male) subjectivity.

It was the racialization of gendered alterity that enabled Europeans to rationalize the slave trade. Europeans re-mythologized Genesis in racial terms, positioning Africans at the bottom of the Great Chain of Being, a metaphysical, "scientific" and sexualized hierarchy at the peak of its acceptance during the eighteenth century (see Wahrman 2004, 131). In their exploitation of alterity through the slave trade, Europeans imagined they were justified by religion and, later, science. Forgotten was the genesis of race in the disavowal of desire. What followed was the structuration of the Other through specularization, rendered rational through scientific observation. Forgotten in the triumph of "ocularcentrism" in the presentism of modernity was the prehistory of race in incestuous desire disavowed.[2]

The ancient Israelite taboo against looking at the naked body of the father represented a ritualized repudiation of incestuous desire. Through the son's eyes the father experienced his nakedness, his vulnerability, his desire, a "lack" Noah denied. Converted to a curse, lack denied became alterity, first gendered, later racialized. In this sequence, gender is the father of race. Racialization becomes formulated through a sexually sublimated specularization: Africans and their descendents were characterized by what Europeans and their descendents "saw" when they "looked" upon the naked bodies of those who had become "Other." It was Noah's revenge; now "he" looks at the naked body of the son, a body safely enslaved, enabling him as the "viewing subject," Kaja Silverman (1988, 5) points out (in a different but relevant context), to protect himself from the "perception of lack by putting a surrogate in place of the absent real. The surrogate becomes the precondition for pleasure."

Freud devised the notion of the primal scene while working with a Russian man named Sergei Pankejev, known as the "Wolf Man." The elements of the primal scene (for Pankejev, they were the wolves he dreamed were staring at him) Freud construed as "products of the imagination, which find their instigation in mature life, which are intended to serve as some kind of symbolic representation [racial representation, I suggest] of real wishes and interests, and which owe their origin to a regressive tendency to an aversion from the problems of the present" (Freud 1963, 236–237; quoted in DiPiero 2002, 36). The "real wish" is the desire of the son for the father and the father's for the son, an incestuous wish made taboo, but which, disavowed, structured racialization. Lynching—the castration of mostly young black men by white men who imagined themselves defending virginal white ladies—is one clue. To make himself "whole," literary critic Lee Edelman (1994b, 65) points out, the white male lyncher violently appropriated the black man's phallus, in the process acknowledging himself to be a "hole" desiring, in the logic of internalization, to be filled.

Social psychologist Roger Brown (see 1965, 751) linked lynching and the miscegenation taboo with incest, missing, however, that the originary biblical act of incest resulting in the "black race" was Ham's defrocking of his father Noah. In order to "understand" this primal scene, DiPiero (2002, 35) points out, we must attend not only to the "manner" in which the "aboriginal event" is "inferred" from its "effects," but to the "reasons" for constructing a primal scene at all. The hegemony of specularity in racialization structures the "manner" of inference of the primal scene of race in the West, and the sexualization of racialization constitutes its "effects."

For me, the reasons for reconstructing the primal scene of race in the West are curricular. What can the study of this primal scene provide us who teach in the present? What can be the pedagogical point of recovering a lost origin, except to enable us to understand more fully whom we have already become? How can we understand the continuing and mutating forms of white racism unless we appreciate that, at its genesis (in the white imagination), it was an incestuous "aboriginal event" between men, a sexual struggle between father and son recoded by subsequent generations as racialized. The father repudiated his son's desire (expressed genitally and/or visually) because it violated his status as a "man," a category that, in its patriarchal formation, requires those who claim it to assert agency, power, possession. As an object of the male son's desire (and we cannot rule out: as the subject of desire for his son), the patriarchal edifice threatens to shatter, stimulating a process of regression to an earlier psychosexual stage in which the infant son, like his sister, is identified with the maternal body.[3]

We can study this psychosexual shattering and its racialization in the infamous case of the late-nineteenth-century German judge Daniel Paul Schreber. Schreber's psychotic breakdown—recorded in his *Memoirs* (1903)—provided Freud with his original theorization of paranoia as an effect of repressed homosexual desire, a theorization no longer taken seriously by most practicing psychoanalysts, but an idea that reverberates loudly in the psychopolitics of whiteness. Like Noah, Schreber claims direct contact with God-the-Father, a contact Schreber finds, against his will, sexually stimulating, and which turns him into a "woman." Like Ham, this son, too, was cursed by God-the-Father, leaving him wounded and enslaved, in Schreber's case, in his hallucinations.

This sexualized and specular structure of white racism fantasizes blackness on the surface of the body, a "colored" surface of skin organized, in the white mind, genitally. In late-nineteenth-century America, white Southerners (and many whites in the North) "saw" young black men as rapists or potential rapists. One hundred years later, and not just in the South, the "rapist" has morphed into the "stud," and black hypermasculinity appeals not only to gay white men, but to white men who imagine themselves "straight" and are also obsessed with black men: with hip-hop performers, with athletes, with the "thug" mystique. Whether rapist or stud, white (including straight male) attention remains fixed on the surface of the black body, and, especially, the (imagined) black phallus. Almost five hundred years later, whites remain mesmerized by the black body, remained deformed by that desire's disavowal: paranoid, predatory, possessive. Black subjectivity remains effaced and, by black critics' accounts (see, for instance, West 1993), too many young black men believe they are primarily their bodies and, specifically, their phallus.

Desire disavowed does not disappear; the repressed "returns" in mutated form, and not only in the fetishization of the black male body. "One of the most consistent medical characterizations of the anatomy of both African American women and lesbians," Siobhan Somerville (2000, 27) points out, "was the myth of an unusually large clitoris."

In late-nineteenth-century Europe, it was a circumcised phallus that provided the fetish for white fantasies of vulnerability, desire, and emasculation. These were fastened to the body of the Jew. (Unfortunately, the Jewish body did not escape the attention of European Americans, as the lynching of Leo Frank testifies.) In contrast to European Americans' hypermasculinization of the African-American male body, Europeans feminized Jews. Like African Americans, Jews were imagined a race apart: sexually rapacious, ethically nefarious, culturally contaminating. As they did in the U.S. version (see Pinar 2001, chapter 6), misogyny and homosexual panic (they are interrelated, of course) structured the European crisis of masculinity, a crisis animating the creative strategies of artists such as Frank Wedekind, Thomas Mann, and Wassily Kandinsky (discussed in chapter 5). It was a crisis theorized theologically and performed sexually by Daniel Paul Schreber.

The curse of Ham becomes "deferred and displaced"[4] in various rituals of servitude, in which the "grandson"—the young male body—is branded by the father, signifying its status as property of the patriarch, a status codified genealogically, subjectively interpellated, and anatomically marked as circumcised. The son not repudiated become genealogical property: a member of the "family." The father's repudiation of his son's desire (and of his own incestuous desire for his son) is signified by circumcision. Unlike Schreber, Noah rejects his own "castration"—as men have tended to characterize negatively what they experience as their "femininity"—by projecting it onto the possessed son, whose penis is then marked to document the interpellating event. Circumcision signifies the sublimated son's "covenant" with his father, with God-the-Father, images of paternity with, as we will see, blurred boundaries.

The father's preoccupation with the son—convoluted as it wavers between the sexual and the sublimated—is not, however, restricted to the West, as studies of coming-of-age rituals worldwide suggest (see chapter 1, "Coming of Age"). Those coming-of-age rituals in which semen is exchanged between older and younger men implies the apparent universality of this sexualized, often sadistic, interest of fathers in sons.[5] Whether "giving head" in New Guinea or "becoming head" of household in the West, the son is relegated to servitude, sublimated or sexualized. Black servitude, of course, was not confined to a specific coming-of-age phase; it was a life sentence without parole: eternal life.

With Thomas DiPiero (2002, 15), I challenge the assumption that (hetero)sexual difference constitutes the "founding" or "fundamental difference" in "human subjectivity." I argue that that self-same sexual difference is the founding or fundamental difference in subjectivity in the West. Its repudiation and projection condemned women and Africans to sometimes conflated positions of sexualized servitude. In Genesis, the splitting off of self-same desire by God-the-Father created (wo)man from the rib of "man," presumably an "opposite sex" who, for men, has tended to function as a displaced and symbolic extension of what he himself is missing. In this patriarchal fantasy, "woman is man minus the phallus" (Grosz 1994, 277). The fundamental difference within male subjectivity in the West is this splitting off of self-same desire, and its consequent abjection as an "abomination," a refusal to know (performed through specularized observation), an insistence on genealogical possession, an obsession to enslave.

In the West, incestuous desire disavowed mutates into epistemology, as Louis Sass' scholarship on Schreber makes concrete, and which David Levin's and Martin Jay's studies of the hegemony of visuality in the West make abstract and general.[6] Louis Sass (1992, 253–254; see Santner 1996, 173–174 n. 35) employs Foucault's notion of panopticism to characterize Schreber's hyperconsciousness, casting Schreber's "rays of God" not as libidinal cathexes but, rather, as "symbolic representations" of Schreber's own "consciousness," a consciousness both "rent" and "joined" by an internalized "panopticism." In this view, the nerves represent those elements of subjectivity that are observed—"self-as-object"—and the rays represent those (especially mental) elements that do the observing, that is, "self-as-subject" (Santner 1996, 174). The God who "lies behind" the rays, Eric Santner (1996, 175) points out, "corresponds" to that "invisible, potentially omniscient," but "only half-internalized Other" who is the "source" and "grounding" of Schreber's specific form of "introversion." It is, as we will see, an introversion that produces "inversion."

Although the hegemony of ocularcentrism in modernity—in particular its political expression as panopticism and surveillance—cannot be attributed directly to that mythic night in Noah's tent, the scholarship does suggest that the disavowal of self-same sexual desire structures alterity through specularity. In the beginning was the word, a "speech act . . . finally less verbal than libidinal," Kaja Silverman (2000, 16) explains, in which self-same sexual desire splits off into "opposite" sexes, a self-division that multiples "others," including "opposite" races. The racialization of alterity through a sexualized specularity produces a paranoid fear of "difference" associated with "others," including the Big Other,[7] but rarely with oneself.

Split off from the self-same body into "opposite" sexes and races, self-same sexual desire is no longer auditory and tactile, but systematically specular, a disembodied, de-individuated mode of visual perception and relation that commodifies and quantifies (see Silverman 2000). The "scientific" systematization of the Europeans' sexualized racial commodification of the Other becomes structured by the epistemology of observation, itself an institutionalization of knowledge production displaced from that interiority self-same difference and desire denied relocated to the exteriority of the bodies of others.

The color of sex, Mason Stokes (2001) asserts, is black. For many whites, the character of "black" is sexual. While the phenomenon of "race" is hardly as simple as that sentence suggests, it cannot, I believe, be grasped or historically surpassed without understanding the relations among alterity, specularity, and the disavowal of incestuous desire between father and son. I am suggesting that the first two follow from the third, that they represent, in part, symbolic wounds of the father (once a son) as he curses the son's son (one day a father, perhaps, who shall carry on the curse). The covenant, requiring filial obedience and generational reproduction, institutionalized the repudiation of that incestuous desire visuality threatened to expose. The injuries it inflicts do not originate in a literal event, of course, but in a mythic one; nor are they contained there. They are restimulated and given aggressive, indeed, vicious social and political form during specific moments in the history of the Western imagination. That vicious social form is the curse, not only that of Ham, but the other curse, the servitude of those sons who sublimate (who do not look, who pretend he is not naked) and who are rewarded with the kingdom of God: that racialized patriarchal system wherein not only women constitute "units of currency" in "gracious" submission to men who imagine themselves white.

How can we educators work to make whiteness conscious of itself, and in so doing, help dissolve it? Teaching for tolerance is not enough. The concept of "citizenship" functions, Russ Castronovo (2001, 212) observes, to "dehistorize historical conditions." Antiracist education cannot be only attitudinal; it must be historical and theoretical. We must theorize the sediments of experience visible today by devising new interpretations of ancient attitudes and practices. Present experience is, as Pier Paolo Pasolini understood, a palimpsest. If "the world . . . was, at first, a pure source of sensations expressed by means of a ratiocinative and precious irrationalism," Pasolini speculated, and "has now become an object of ideological, if not philosophical, awareness," then, "as such, it demands stylistic experiments of a radically new type" (quoted in

Greene 1990, 37). I propose a radical revision of the synoptic text (2004d; in press). For those teachers who appreciate the centrality of academic knowledge in the cultivation of self-reflexive and ethical intelligence, I provide summaries of scholarship the juxtaposition of which might make those civilizational sediments discernible. We might gain what Lee Edelman (1994a, 268) calls "(be)hindsight."

I employ a version of what Dror Wahrman (2004, 47) characterizes as "repetition in the second degree," in which I focus on specific "cultural domains that push the argument and its limits in variously revealing ways." The first domain is ancient Israelite culture, namely Noah and the so-called curse of Ham, in which racialization is retroactively realized; the second is the crisis of late-nineteenth-century European masculinity and the case of Daniel Paul Schreber, including Freud's engagement with it; the third is the history and culture of circumcision, the mark of the covenant; the fourth is coming-of-age rituals in the South Pacific, the cultural complements to circumcision in which the sexual side of the sacred is exposed; and the fifth is the traces of these four domains in contemporary representations of race in literary and popular culture. Juxtaposing these five domains creates a pattern—a collage, however weak—that points to the incestuous genealogy of whiteness and of the racism it requires.

Despite my pedagogical good intentions, this strategy recalls what Wahrman (2004, 44) identifies as the "most problematic methodological quagmire of cultural history, which he describes as the "difficulty" of the "weak collage." Identifying "seemingly similar phenomena in several disparate cultural spheres at the same historical moment," Wahrman (2004, 44) worries, the historian declares them a "pattern of historical significance." My claim here is that the pattern this textbook reveals holds *educational* significance, but the dangers of associating different phenomena in disparate cultural spheres during different historical moments are, I trust, obvious. Serious students will return to the original texts themselves in order to study the singular elements comprising this synoptic collage.

Acknowledgments

In 1985 I moved from Rochester, New York, to Baton Rouge, Louisiana. This book derives from my experience in Louisiana, including my adventures in New Orleans' Old Quarter and conversation with colleagues and students at Louisiana State University.

Fifteen years ago LSU literary theorist Rick Moreland and political scientist Wayne Parent generously engaged me after my initial efforts to understand my experience of the South; this work started then.

This book occupied me during my final five years at LSU, during which colleagues and students listened patiently to various fragments from it. Among these were my colleagues associated with LSU's Curriculum Theory Project: Bill Doll, Petra Munro, Denise Egéa-Kuehne, Nina Asher, Nancy Nelson, Claudia Eppert, Kaustuv Roy, Becky Ropers-Huilman, Abul Pitre, and, most recently, M. Jayne Fleener. Among the Ph.D. students who engaged me were Nicholas Ng-A-Fook, Donna Trueit, Nichole Guillory, Brian Casemore, Ugena Whitlock, Laura Jewett, Sean Buckreis, and Hillary Procknow. My colleagues Ron Good, Ann Trousdale, and Jim Wandersee also contributed to the extraordinary scene that has been LSU's Department of Curriculum and Instruction.

Outside LSU, my special thanks go to Alan Block, who generously helped me grapple with Noah.

My LSU colleagues in Women's and Gender Studies have been important to me as well. In particular, I thank Barbara Apostelou, Katherine Powell, Michelle Massé, Joyce Jackson, James Catano, Katherine Henninger, Jennifer Jones-Cavanaugh, Gail and Peter Sutherland, and, especially, Elsie Michie. While many have contributed to this thought experiment, no one has helped me more than Elsie Michie. I thank you, Elsie, for your intellectual encouragement and for your friendship. I shall miss those lunches at Kamado's.

Introduction

[T]he racial is itself sexualized in U.S. culture. (Robyn Wiegman 1995, 162)

Race and ethnicity are thus coterminous with sexuality, just as sexuality is implicated in race and ethnicity. (Rey Chow 2002, 7)

[G]ender is the modality in which race is lived. (Paul Gilroy 1993, 85)

[T]he acquisition of gendered identity in liberal capitalist societies is always a racial acquisition. (David Eng 2001, 17)

Whiteness itself is predominantly a male franchise. (Mason Stokes 2001, 46)

That night in Noah's tent may be the genesis of "race" in the West, but there are other originary moments: Genesis 1:27, when "man" is created first in the image of God, a male God, and Genesis 2:21 and 2:22, when "woman" is fashioned from the "rib" of "man" (leaving a wound that will be reopened at the crucifixion of Jesus.) These gendered and originary fantasies are deeply imprinted in the Western imagination, expressed in the "one-sex" theory of humankind (see Laqueur 1990), rendering "woman" a displaced element of "man," relegating "woman" to what men have imagined "her" to be. "Her" displacement to an apparently anatomical Other creates the binary— man/woman—by means of which is perpetuated the illusion of opposite sexes, setting the gendered stage for opposite races.

The experience of this wound, in psychoanalytic terms, is an experience of lack or "castration." Teresa de Lauretis, who asserts we cannot think the sexual outside psychoanalytic categories (see 1994, 30), links castration to dispossession, due to its structural role in the formation of a fetish. Recall that in Freud's account the fetish is a substitute for the "maternal penis" that the male child expects to see on the mother's body but finds absent; his sighting of a body without penis produces, presumably, castration anxiety, later relieved by his fashioning of a fetish.

In Freud's speculation there is a refusal, as Kaja Silverman (1988, 15) points out, to associate castration with "any of the divisions which occur prior to the registration of sexual difference." This refusal, she suggests, discloses Freud's insistence on "distance between the male subject and the notion of lack" (1988, 15). To acknowledge that the loss of the object—for instance, the loss of the maternal body that birth inaugurates—constitutes a castration would be to appreciate that the male subject is "already structured by absence prior to the moment at which he registers woman's anatomical difference" (Silverman 1988, 15). Like Eve, then, Adam has already been deprived of self-identical being; he is already "marked by the language and desires of the Other" (Silverman 1988, 15). The name Adam means "clay," and it was out of the dust of the ground that the Father blew life into the "vessel" that was the first "man" (Mitchell 1994, 167). In this patriarchal hallucination, his companion is fashioned from his own body, a displacement of otherness from the self-same body. S/he is more accurately named (St)Eve.

Fetishism does not tend to occur among women, de Lauretis suggests, since anxiety over castration is produced by the daughter seeing a body with penis, a male body, and not the mother's body, which is, presumably, "like" her own. For daughters and sons, de Lauretis suggests, the fetish enables the disavowal of lack or loss by representing the object that is missing but narcissistically wished for: the penis-phallus in one case, the female body in the other. For each, "the fetish, in its various contingent forms, is nothing but a signifier of desire" (de Lauretis 1994, 308).

For David Eng (2001, 5), "castration is always racial castration." From its inception, he argues, "psychoanalysis has systematically encoded race as a question of sexual development" (2001, 6). He points to Freud's treatise on the relationship between "primitive" sexual practices and "civilized" neuroses as indicative of Freud's "discursive strategy" (2001, 6). Eng (2001, 6) notes that Freud opens *Totem and Taboo* by focusing on the figure of the "primitive" in order to trace the "dark origins" of the contemporary European psyche. "There are," Freud wrote, men still

> living who, as we believe, stand very near to primitive man, far nearer than we do, and whom we therefore regard as his direct heirs and representatives. Such is our view of those whom we describe as savages or half-savages; and their mental life must have a peculiar interest for us if we are right in seeing in it a well-preserved picture of an early stage of our own development. (Quoted in Eng 2001, 6–7)

Would he confer upon the Sambia such originary and primal status? (see chapter 1, "Coming of Age")

Because these gendered and racialized positions are binary, they are inherently unstable. Precisely because they are contingent upon the relation of each to the other, altering the gender or race of the other threatens (but, unfortunately, does not destroy) the entire binary structure. That ancient Israelite men were commanded to love a male God threatened to feminize them, as Howard Eilberg-Schwartz (1994) makes clear. Circumcision was one fetishistic strategy by means of which men loving their God-the-Father–phallus could perform that desire by signifying its sublimation. No such games for Schreber: he understood that loving and being loved by the Father meant his own emasculation. For this distinguished judge and son of a famous educator, that meant madness.

In Protestantism, male "brides of Christ" also would experience their masculinity as imperiled, as historian Philip Greven (1977) has documented. The trick was to remain the "active" member in the pair, by loving God with all one's might and thereby identifying with him, imagining that one speaks in his name. But the trick fails to fool, as the self-effacing love of the supplicant invites an "active" God whose desire renders men as "women." In the sexualized presence of the Almighty, that "rib" flies back into the man's now-swollen chest, swollen not with muscle, but with breasts, the very same swollen breasts he demands women to exhibit. Noah inverted becomes Daniel Paul Schreber. None of this is obvious to students, especially white students in the neo-Confederate South, coming of age in that masculinist and predatory America George W. Bush summons and personifies.

African Americans or other "people of color," Kalpana Seshadri-Crooks (1998) points out, often speak of their awareness of "race" as a conscious, historical discovery. In contrast, whiteness is rarely recalled or "seen." Mason Stokes (2001, 185) concurs: "Let's be honest: white people have never been very good at thinking about their own whiteness." This fact points to the "deep relation" between "whiteness" and the "unconscious" (Seshadri-Crooks, 1998, 358). In a heterosexist West, this reference to a cultural unconscious suggests not only whiteness, but homosexuality and the "negative" Oedipus complex as well. The white unconscious is queer and black.

This unconscious content is evident in Stokes' (2001, 166) observation that the use of white actors in blackface in Griffith's *Rebirth of a Nation* precipitates the "ultimate collapse of the system of binary differences that Griffith longs to maintain." Stokes quotes Rogin to underscore his point: "The obviousness of blackface, which fails to disguise, reveals that the Klansmen were chasing their own negative identities, their own shadow sides" (1985, 181; quoted in Stokes 2001, 166). "The climax of *Rebirth*," Rogin writes, "does not pit whites against black, but some white actors against others" (1985, 181; quoted in Stokes 2001, 166). That racism is about a split-off white masculinity comes as no consolation to black victims for whom the contents of the white unconscious have had, and continue to have, horrific material effects.

Whiteness can be said to function as a structural condition of dominant subjectivity, Seshadri-Crooks suggests; it inserts the subject into the symbolic order. Such insertion requires self-splitting, in which language inscribes in the subject an identity that is surface and representational, in the process estranging the white male subject from its "self" and engendering the un-conscious as "feminine" and "queer," racializing it as "black." By its insertion in the symbolic order, the son becomes gendered and raced as "white," constituted as a speaking subject (Seshadri-Crooks 1998). He claims to speak for God-the-Father; he can curse his son and in so doing, he names reality.

This structural specificity of whiteness underlines the inextricable relation between sexual and racial identity. For Seshadri-Crooks, this means we must discern how the seemingly extrafamilial signifier of race, which critics consign only to the public sphere since it seems to conjure up a collectivity, intersects with that of sex to produce a private racialized subject who is hypersexualized. This intersection suggests, Seshadri-Crooks (1998, 258) continues, that we "reread theories of sexual identification in order to render visible the color of the law enforcing racial and sexual normativity at perhaps the same moment." In the West, heteronormativity and whiteness are conjoined (Stokes 2001).

Seshadri-Crooks' theorization of the unconscious character of whiteness and its successful reiteration turns on the fact that identity is always partial and subject to failure and, Petra Munro (1998) would add, "subject to fiction." This failure occurs because, like sexual identity, the constitution of "race" does not occur as a complete "event." After Judith Butler, Seshadri-Crooks notes that the process of identification is performative. The process is never finally achieved; it is always seduced by the desire from which it flees. In the structuration of alterity is always the "I," the displaced rib. Shem and Japheth—the sublimated sons—choose identification over desire. It is the choice enabling what Kaja Silverman (1988, 215) terms a "secondary" identification, "leading to imaginary mastery and transcendence."

The identificatory split between the "I" and the "we," structurally analogous to the subject's self-splitting in language (between the I who speaks and the I who is the subject of speech), splits whiteness into a consciousness of being white and a desire to possess whiteness—to be not black, not queer. This self-splitting structures, makes defensive and compensatory, hegemonic masculinity, in particular, white straight masculinity, no monolithic subject position to be sure (Pfeil 1995). It is this internal rupture that Seshadri-Crooks, after David Roediger, suggests may be a nodal point, a structural point of vulnerability on which to focus a curriculum for studying whiteness and, necessarily (given their inextricable interrelationship), heteronormativity. Any curriculum that focuses on the one without critical attention to the other is camped outside the tent. I am suggesting we whites must go back inside the tent and meet our maker.

Seshadri-Crooks draws a parallel between gender and racial identification, quoting Butler's important point that the abjection of homosexuality enables, even produces, heterosexuality. Butler's argument that the disavowal of homosexual desire founds heterosexuality relies on the view that desire and identification are not mutually exclusive, that the identificatory recovery of the sexually prohibited object constitutes the gendered construction of sameness. In other words, identification functions to produce "homosociality," that is, affiliation and solidarity, a racialized fraternal fascism.

While claiming racial sameness has tended to result in claims of universalism, Seshadri-Crooks notes that it has also tended to obliterate racial difference and reproduce (indeed, procreate) racial hegemony. Seshadri-Crooks argues that whiteness requires the abjection not of an identification with color per se, but of *whiteness* as a color. Desire denied does not disappear; it mutates, it is deferred and displaced. Fetishism follows. As in the case of lynching and prison rape in the United States, the denial of the specificity and contingency of whiteness coupled with an ongoing "crisis" of white masculinity functioned to fetishize the black phallus (see Pinar 2001, chapter 19, section III).

As Elisabeth Young-Bruehl (1996) points out, Frantz Fanon (1967) took note of the inability of whites to appreciate how whiteness informs their psychological, and, specifically, racialized experience. For the "Negro," Fanon argued, the experience of race is never hidden or repressed. Racial drama is always visible, always conscious. Because it is conscious (and not unconscious, as it is for whites), Fanon insisted, the nightmare of racism does not make blacks neurotic, even if it often deposits a (t)race of internalized inferiority. Seshadri-Crooks suggests that African Americans can tell—or can express in an artistic medium—their stories.

For whites, the drama of race is rather different; it is unconscious, evident in some forms of psychoanalysis, in which (white) people are, presumably, first shaped by their families, and later and secondarily by encounters with the society and culture in which their families are embedded. But families are not havens in a heartless world (Lasch 1977); they are complex, if singular, configurations of that world. Society and culture are internalized and reproduced through the family.

For Fanon (1967), it is white people who are structured by the Oedipus complex, which, Fanon argued, does not exist among blacks in the French Antilles in a form comparable to that described by Freud. Moreover, whites are prone to phobia, a condition following from trauma. That trauma, for Freud, is the anger-inducing displacement of the mother or the loss of her to a sibling or paternal rival, or, perhaps, to one's (male) self. In terms of the male's "negative" oedipal trauma, this trauma is also the loss of the father, in heterosexist culture less in the rivalry of others than to the internalization of the heterosexist imperative: desire is converted to identification.[1]

Fanon was describing only the former trauma, but both forms work in his argument: namely that to these losses accrue secondary traumas associated with more-or-less imaginary attackers or sexual abusers. Whites are destined to lose their loved ones, Fanon argues, and in ways which recapitulate the first oedipal losses. Phobia-prone white people, Fanon (1967) suggests, fear sexual loss, theft, and assault. More specifically, both white men and white women fear black men: "the Negrophobic woman is in fact nothing but a putative sexual partner—just as the Negrophobic man is a repressed homosexual" (Fanon 1967, 156). Significantly, Fanon (1967) equated racism with masochism (see Young-Bruehl 1996, 497). Although he tended to focus upon the white woman, the formulation works for white men as well. Perhaps Noah was not raped by Ham; perhaps the son was raped by the father. The "curse" was, in this sense, completed by the perfect crime; the father blamed the victim.

What emerges from the tent, then, is a "sexual structuring" of alterity, a gerund-phrase Teresa de Lauretis (1994) prefers to more familiar ones like sexuality or sexual identity because the gerund form conveys, both etymologically and performatively, an ongoing activity, an interactive process. De Lauretis points out that *gerund* derives from the Latin *gerere*, to carry, and the gerund form communicates the dynamic meaning of the verb, something not conveyed in terms such as *identity* or *sexuality*. By using the term *sexual structuring* de Lauretis underlines the constructedness of sex and of the sexual subject. The effects of this ongoing historical and cultural process do not derive, she points out, from an originary materiality of the body, nor do they modify or attach themselves to a gendered or sexual essence—either corporeal or existential—prior to the process of structuring itself. "[N]either the body nor the subject," de Lauretis (1994, 301–302) explains, "is prior to the process of sexuation; both come into being in that continuous and life-long process in which the subject is, as it were, permanently under construction."

In the racialization of alterity, specific forms of sexual structuring seem especially germane. Mason Stokes (2001, 14) claims that the "chief ally in whiteness' normalizing vision is heterosexuality." He cites Monique Wittig's definition of heterosexuality: it is

> a nonexistent object, a fetish, an ideological form which cannot be grasped in reality, except through its effects, whose existence lies in the mind of people, but in a way that affects their whole life, the way they act, the way they move, the way they think. So we are dealing with an object imaginary and real. (1992, 40–41; quoted in Stokes 2001, 14)

In the context of masculinity, as we will see, it becomes an unattainable ideal.

Like whiteness, heterosexuality's power to promote its own invisibility is so complete that "the straight mind cannot conceive of a culture, a society where heterosexuality would not order not only all human relationships but also its very production of concepts and all the processes which escape consciousness, as well" (Wittig 1992, 28; quoted in Stokes 2001, 14). Those "processes which escape consciousness," Stokes (2001, 14) notes, constitute the "primary breeding ground for heterosexuality's normalizing and self-generating power." Significantly, Stokes (2001, 14) adds: "the same holds true for whiteness."

Invisibility would seem a strange liability for a culture mesmerized by the eye. There are, critics of ocularcentrism tell us (see, for instance, Levin 1993b), key moments in the West's epistemological privileging of vision, among them (a) Plato's notion that ethical universals must be accessible to "the mind's eye"; (b) the invention of printing; (c) the appearance of the modern sciences; and (d) the rise of "race." Descartes saw truth in clear and distinct ideas requiring a "steadfast mental gaze," while Bacon posited observation as the prerequisite for objectivity, linking knowledge itself with sight. Such a privileging of the visual has hardly gone unnoticed, as evident in the work of late-nineteenth- and early-twentieth-century philosophers such as Henri Bergson, Friedrich Nietzsche, and Martin Heidegger (Levin 1993b).[2]

In our time, a philosophical suspicion of vision is not only evident in Foucault's analysis of power and, specifically, surveillance, but in a series of philosophical critiques, not only (but, it seems, especially) in France but in America as well (for instance, in Dewey's critique of spectator knowledge and in Richard Rorty's critique of ocular metaphysics in Western philosophy).

These critiques expose the epistemological and political privileges usurped by ocularcentric technologies of interpretation codified as observational protocols presumably scientific in nature. In its fascination with external appearances and observable behavior, these critiques suggest, sight overlooks subjectivity and inner meaning. It tends to be presentistic and, thereby, ahistorical. In contrast, interpretation makes meaning audible by listening to "complicated conversation" (see Pinar 2004a, 9; see also Pinar and Irwin 2005). Such auditory opportunities are what contemporary curriculum research aspires to provide, in the present instance, curricular opportunities for teachers and students to dispel the curse of Ham.

Noah and Schreber are asymmetrical figures in Western civilization's specular racialization of alterity. Naked, Noah disavows his lack and curses his progeny to servitude—genealogical property branded by circumcision. Schreber succumbs to his lack; the son is seduced by the father, property still. Noah re-robes himself as patriarch; Schreber dresses as the woman his heavenly Father wants him to be. In its denial of the maternal body, male subjectivity is banished to the surface of the male body, from where men project—"see"—reality, engendered and racialized.

Whether Ham's transgression was the sexual penetration of his father or "merely" looking at the naked body of the father, in both instances the son saw his father's "lack," an embodied state of "castration," denied in the curse. Lack denied displaces alterity from within, from the self-same body, and projects it onto an "other" imagined as different due to anatomy. Our work as antiracist educators requires returning to Noah's tent. If we are once again our father's lovers, might we sons become our brother's keepers and our sister's friends?

Unlike Robert Musil, I am not recommending incest as utopian. Musil's fantasy, in *The Man Without Qualities*, was heterosexual incest between brother and sister. For me, Noah's tent is a metaphoric formulation of the white male "self." To re-enter it is to "regress" toward a re-experiencing of what Freud characterized as the "negative" oedipal complex, and a subsequent restructuring of internal object relations in which binaries are mixed and merged in what becomes the self-same (but now simultaneously the "opposite" sexed) body. To become "race traitors," then, we white men must become gender traitors, although, as Robyn Wiegman understands, to enable reparation, not redemption.

Reinserting Adam's rib will not reset the matrix. Still, we must re-experience our subjective relation to the internalized father not as dis-avowed, competitive, and contentious (as "cursed"), but as symbiotic and "incestuous" (homosexual not homosocial). In such an enfleshed restructur-ing of the inner topography of white masculinity, we might welcome the opposite sex and race within our own psyche (see Geyer-Ryan 1996, 123). Then, perhaps, the curse that is "race" can be subverted. There are reparations to pay, and not only subjective ones. Let us study the civilizational legacy that has left us so indebted.

CHAPTER 1

In the Beginning

THE NOAH COMPLEX

We shall see how the emergence of an idea of race is inextricably linked to an increasing preference for these profane stories—the legend of Noah, the curse of Ham, the mark of Cain—as literal explanations for the origins and divisions of man. (Ivan Hannaford 1996, 133–134)

By staring at a complicated passage in Genesis while experiencing human difference in racial terms, readers may have projected some of their most feared transgressions onto Noah's sons, both as a cause or an explanation and as the result of difference. (Werner Sollors 1997, 73)

The black man was there, in the beginning. (Mason Stokes 2001, 107)

[I]t was as if single nights had the duration of centuries. (Daniel Paul Schreber 2000 [1903], 76)

However imaginary Noah may be, "he"—like "Adam"—is "a figure we can discern to be characteristically male and patriarchal, but which, if it were not for those human attributes, would rise above our earthly existence and join the ranks of the demi-gods" (Mottram 1937, 2–3). While Adam was the father of the first race of man, Noah is regarded the father of the second. So understood, Noah is the second Adam; in him, presumably, God cast the future of the race (Cohen 1974). Both constitute the main characters in a "fantasy of masculine autogenesis" in which the "maternal origin" remains "unthematizable materiality" (Vasseleu 1996, 130).

God's promise to Noah to lift his spell from the earth, thereby guaranteeing the continuation of life, performed the prediction made by Lamech that Noah's father, when he named his son "Noah," a word corresponding in sound if not in origin, "will provide us relief" (Cohen 1974, 56). Lamech correctly foretold that "this one will provide us relief from our work and from the toil of our hands, out of the very soil which the Lord placed under a curse" (Gen. 5:29). Is this a promise to provide slaves? Is what happened in the tent a setup, a script composed by God-the-Father for his son Noah to follow, enslaving his son's son, thereby providing "relief from our work and the toil of our hands"?

Genesis tells us nothing about Noah's brothers and sisters. Noah's sons—Shem, Ham and Japheth—have, then, no uncles or aunts or cousins. Noah was born to his father at biblical "middle age" (at one hundred and eighty-two, presumably); Noah himself became a parent at five hundred. "Was it a naïve desire to raise man to the dignity of the Mammoth," R.H. Mottram (1937, 11) asks, "that caused the insertion of those fantastic figures?" Not only those figures are "fantastic," of course: in Genesis 9:24, the Bible becomes a fantasy of incestuous desire denied, desire displaced through a curse, restructured as servitude.

Ham is the son of Noah whose name is generally recorded first and who may have been, although not the eldest, perhaps in status the senior of the sons. It was Ham who went to the tent for a reason we are not told. It is not even clear, Mottram observes, if we are to assume that the family was still living together as a family, with Noah as its patriarch and exercising daily authority. It was inside Noah's tent, of course, that Ham looked at his father, "revered, heroic" (Mottram 1937, 188), lying naked and "dead drunk." Ham then left the tent and told his brothers, who "took a garment, and laid it upon their shoulders and went backward and covered the nakedness of their father; and their faces were backward and they saw not their father's nakedness." Was their action stimulated by their reverence for the heroic figure? Were they only doing for their father what he had neglected to do for himself? There is no sexual hypothesis in Mottram's account: "They covered him decently. They did not wake him" (1937, 188).

Noah slept it off. When he "awoke from his wine," someone (who?) told him what had happened, or so Mottram suggests. Or did he know himself? The narrative only reports: "He knew what his younger son had done unto him." Declining a sexual interpretation, Mottram tells us that Noah knew "that he had been made ridiculous" (1937, 191). He had been *made* to look ridiculous? By himself? Is this a bourgeois projection? More interesting, Mottram speculates that "what really happened was that he became self-conscious" (1937, 191). Is anal penetration a prerequisite, even a metaphoric one, for men becoming conscious? Returning to his bourgeois preoccupation, Mottram speculates that Noah concluded "that he had a dignity to maintain, and that it had been treated lightly" (Mottram 1937, 191). "Cursed be Canaan," he cried; "a servant of servants shall he be to his brethren." He went on: "Blessed be the Lord God of Shem and Canaan shall be his servant." He added: "God shall enlarge Japheth and he shall dwell in the tents of Shem, and Canaan shall be his servant."

Mottram notes that these lines—in which the "curse of Ham" is pronounced—may well have been written long after the event. They may have been written when the disparity between the fortunes of the sons of Noah (or those tribes which had become represented by those historic names) had to be explained. In this scenario, the writer created the "curse" to characterize the more successful and prosperous as the more devout descendants; *they* had acted in some more properly filial manner toward that legendary figure from whom they all came (Mottram 1937). The name of

Japheth suggests that very "enlargement" that was promised to him; that of Ham, means "dark"; and Shem suggests "renown" (Mottram 1937, 192). Given the ancient and continuing association of "dark" with "evil" and with "dark-skinned" peoples (see Jay 1993, 509), we have in Ham's name the genesis of the racialization of the gendered curse of Ham.

Mottram's account is meant to introduce Noah to the general public. It raises more questions than it answers, questions biblical scholars have studied in detail. H. Hirsch Cohen (1974) suggests that the interpretations of many scholars are inadequate because they have never explained satisfactorily why the one man thought worthy enough by God to be saved from the waters of the flood should later be portrayed later as drunk, naked, and passed out in a stupor. Although Cohen's account also fails to explain the curse, it does provide details that an introductory account by definition does not. Like Mottram, Cohen tells us that Noah's nakedness was directly related to his drunkenness; he believes that both Noah's drunkenness and his nakedness constitute an integral sequel to the flood story.

The *Zohar*, that medieval source book of Jewish mysticism, depicted Noah as having been driven into a drunken stupor by his idealism. The grapevine he had planted, from which the wine he drank was made, had come, he believed, from the garden of Eden. Presumably, he wanted to drink of the vine in order to better understand the sin of Adam, in hopes he might then warn the world. Early church fathers Origen and Chrysostom excused Noah for a different reason. They suspected the old man had not known what wine could do. Drinking alcohol without knowing its effect, Noah quickly became overwhelmed in the inebriating power of the fermented grape (Cohen 1974).

Cohen (1974, 1) characterizes Michelangelo's fresco of the drunken Noah in the Sistine Chapel as the "tragic confrontation" between "youth" and "old age." He points to the "listless and aging body" of the reclining Noah as symbolizing the "infirmity" and "weakness" of "age." The "athletic" bodies of his sons represent the incarnation of youth in its prime (Cohen 1974, 1). To the homosexual Michelangelo, these were, no doubt, more than mere symbols of youth in its prime. Moreover, the erotics of the relationship between Ham and Noah may have been obvious to him. They would have been to Freud.

The fetishistic nature of this myth installs the relation between father and son as the archaic structuration of authority, power, and subordination. The Noah myth reiterates difference as that which separates fathers and sons, the difference between manhood and emasculation, the latter the gender starting point of all men, from which boys must flee to become "men." Father–son incest and its phobic rejection inseminate homosocial society. The structural anteriority of male homosexuality, Trevor Hope (1994) points out, hardly confers any privilege upon it. Quite to the contrary, its status as a primal scene functions to pathologize this archaic, retrojected, homo-sexual/social economy that is specularly structured as "race" and "gender" (see Braidotti 1994b).

What is the meaning of homosexuality generally, and father–son incest specifically, when they are installed as a "repressed historical horizon" to be "recovered" (Hope 1994, 189)? For Hope, such an interpretive calling fails to question the narrative function of the symptom within the disciplinary cognitive regime of the Enlightenment, the historical period during which the epistemology of "race" becomes "scientific." For Hope (1994, 189), it is a "certain obscurity" within the "symptom" that functions as the "eroticized" ground for a "sociality" cohering around a "retrojectively disavowed libidinality." The sociality thereby produced is, of course, Sedgwick's (1985) homosociality and its universalization of the masculine subject position. But what is this "certain obscurity"? Is it father–son incest?

Racialized, this "disavowed libidinality" enables the fantasy of "whiteness," a "race" that is, as Mason Stokes has pointed out, heterosexual. The homosexual relation between God-the-Father and his creation, Adam, and God's son, Jesus, procreates a set of abject "others," among whom the "second sex" (de Beauvoir 1974 [1952]) and, relatedly, the "Negro"—the "lady of the races" (Park 1950, 280)—are prominent. This peculiar disavowal of desire constructs a parallel asymmetry between "opposite" sexes and the radically different "races." The price men pay, Rose Braidotti (1994b, 202) suggests, is an "abstract virility," accompanied by a form of disembodiment (see Pinar 2004c). Is this expressed epistemologically as "science"? The price women (and, I would add, African Americans) pay, Braidotti (1994b, 202) asserts, is an interpellated over-identification "with their embodied condition and especially with sexuality. The relationship between the two poles is postulated in terms of compulsory heterosexuality" and in terms of "blackness."

The striking point Hope makes is that such a "disavowed cognition"—one that establishes disavowal as its epistemological basis—is "fetishistic," according to David Eng (2001, 146), the "paradigmatic example" of "divided belief." Such cognition produces not conclusions but substitutes, powerfully cathected to the displaced and deferred contents of originary desire, which then circulate as violent effects of the primal repudiation. Although mythic in that its origins are "retrojected" into prehistory, this fetishistic cognition of substitution becomes institutionalized during the Enlightenment, the period during which the scientific study of "race" is undertaken. Hope (1994, 189) argues that the epistemology of the Enlightenment rests on a "fetishistic economy." The injunction to "Know!" is animated, Hope seems to be suggesting, by the search for a "lost" homosexuality, lost first, I speculate, in Genesis 1 and, again in Genesis 9. Hope (see 1994, 189) offers that such invocations of male homosexuality represent gestures of distancing and veiling. Is it possible to study these gestures so as to bring them close, unveiled, indeed naked, perhaps in a tent? Is it possible—in imaginative and scholarly terms—to re-enter the primal scene of race in the West?

Cohen acknowledges the considerable interpretive energy that has been expended by biblical exegetes trying to figure out what happened to the drunken, naked Noah inside the tent, "whetted by the enigmatic statement" (Cohen 1974, 13) that Noah, upon recovering consciousness, "knew what his youngest son had done unto him" (Gen. 9:24). "Apparently," Cohen (1974, 13) writes, "more than Ham's voyeurism is involved," but "precisely" what is, is left unsaid by the "narrator." (The love that dare not speak its name will never occur to Cohen, at least not in print.) Due to the absence of details, Cohen speculates, ancient Jewish commentators were compelled to embellish the story with the "lurid details" obviously left out by the narrator (Cohen 1974, 13). These are the "lurid details" of men's fantasy lives.

The rabbinic sages of the Midrash and Talmud concluded that Noah must have been castrated in the tent. Symbolically, the great patriarch had been turned into a woman—a "punk," in prison parlance—not unlike Daniel Paul Schreber. In reconstructing the incident, Cohen tells us, several rabbis imagined Canaan, Ham's small son, entering the tent, looping a cord around his grandfather's exposed testicles, drawing it tight until the deed was done: castration. There is no discussion here of the sages' fantasy that a little boy could (or would want to) castrate the patriarch.

Informing Shem and Japheth of this "gruesome deed," Cohen (1974, 13) continues, rabbinic sages then imagined Ham responding to his father's castration as if it were humorous. Other rabbis, Cohen reports, maintained that Ham, not Canaan, castrated his father, causing Noah to cry out: "Now I cannot beget the fourth son whose children I would have ordered to serve you and your brothers! Therefore it must be Canaan, your first-born, whom they enslave" (Graves and Patai, 1964, 121; quoted in Cohen 1974, 13). Incapable of paternity, I am no longer a "man." I am a "degenerate."

Other sages denied that Ham was guilty of such insurrection, suggesting instead that maybe a lion castrated Noah. (This speculation brings to mind a poster common in gay bars a decade or so ago of a male lion on top of a white man, penetrating him anally. This is, evidently, a persisting fantasy.) Cohen assures us that the sages' imagination was hardly that lurid. They imagined, instead, an accident. As Noah disembarked from the ark, a lion—ill perhaps from the voyage—struck Noah's genitals with an inadvertent swipe of his paw so that he never again could perform the marital act (Cohen 1974). In this version the tent and the wine are irrelevant.

Howard Eilberg-Schwartz allows that those who argue Ham committed a homosexual *act* with his father are "partially right" (1994, 96). There is, he continues, "homoeroticism" in the Noah narrative (1994, 96). Ham's act of looking upon his father's naked body was, Eilberg-Schwartz (1994, 96) suggests, "enough" to stimulate "desire." That "the gaze" and desire are "intertwined" has been, he judges, "amply demonstrated" (1994, 96) by recent art and film critics (among them Norman Bryson [1983]). Eilberg-Schwartz cites John Berger's famous elaboration of the relations among the gaze, power, and heterosexual desire in European paintings of the nude:

Men act and women appear . . . . Men look at women. Women watch themselves being looked at. This determines not only most relations between men and women but also the relation of women to themselves. The surveyor of woman in herself is male: the surveyed female. Thus she turns herself into an object—and most particularly an object of vision: a sight. (1972, 47, 54; quoted in Eilberg-Schwartz 1994, 96)

This familiar passage becomes strange as it foreshadows, in this context, not only the curse but Schreber's seduction.

Eilberg-Schwartz also cites Laura Mulvey's argument that, in film, it is the male heterosexual gaze that directs the view of the camera. The film's viewers are thereby positioned to gaze upon women as objects of desire. "The male figure cannot bear the burden of sexual objection" (1989, 20; quoted in Eilberg-Schwartz 1994, 96). This seems to be changing, if many young men's apparent eagerness to strip is any indication. Young men's recasting of themselves (and/or by heterosexual women and gay men) as "boy-chicks" does not necessarily refute Doane's (1982) observation, which Eilberg-Schwartz also cites, that many consider a female–film character taking off her glasses in films as erotic, presumably a symbolic act in which she is relinquishing her position as a spectator and becoming instead the object of the gaze (see also Silverman 1988).

The male gaze and desire were strongly associated in the ancient Israelite imagination, Eilberg-Schwartz tells us. In general, the gaze that beholds is that of a man looking at a woman. On occasion it is a woman looking at a man, but it is only rarely a man gazing at another man. Given this ancient prohibition against the male–male gaze, the curse of Ham functions to direct the male gaze away from the male to the female body. Given this visual–sexual logic, it was Noah's passivity, his assumption of the feminized position, that made the son's viewing of his naked body so unacceptable. No object, he, in another's man gaze; father must be master. Is this patriarchy's primal scene?

The Primal Scene

The primal scene is always a scene that is "unknown" and "forgotten." (Ned Lukacher 1986, 27)

The unconscious is this: that persistence on another scene, contrary to our clear and distinct reflections, of a link which can no longer be conceptualized. (Jean-Joseph Goux 1992, 52)

By the time "woman" arrives at man's side . . . the coupling of "man–woman" is already obsolete, not so much because its twosomeness is heterosexist as because such a twosomeness itself will have to be recognized as part of something else. (Rey Chow 2002, 160)

"[R]ace" actually contains a host of other social indicators, wrapped up together in a historically produced series that deletes the specificity of each constituent part . . . . [T]hat is precisely where the category of "race" achieves so much of its social power. (Thomas DiPiero 2002, 51)

Webster's New Collegiate Dictionary defines "primal" as, first, "original" or "primitive," and, second, "first in importance: fundamental." Each of these seems to be at work in Freud's (1913) theorization of the concept in *Totem and Taboo*. As Naomi Greene (1990) points out, Freud imagined himself reaching back into prehistory to reconstruct the link between contemporary social rituals and what he imagined as the primordial social scene which inaugurates them. That scene was one of murder, a murder in which the primal father was killed by his sons. The positive Oedipus complex—the son's rivalry with the father for his love of his mother—derives from this primal scene. Such rivalry hardly drives the son and the father apart; indeed, the bond between the two rivals is "stronger" and "more defining" than the bond between lover and beloved (Van Alphen 1996, 172–173).

After the patricide, the sons become plagued by a mix of joy, fear, and guilt, prompting them to (re)incarnate the dead father as a totem animal. Through its sacrifice, performed ceremonially, the sons seek appeasement from the father, who in death has become to them even more powerful. They also seek, through their claims of intimacy with the invisible Father, their own empowerment, as the history of priests and other religious leaders suggests. Levi R. Bryant (2004, 342) argues that

> [T]he democracy of brothers described in Freud's *Totem and Taboo* can only flourish so long as the murder of the primordial father remains subtracted from the discourse as a shadowy memory not to be spoken. It is only on the basis of the shared guilt founded upon this act of murder that the democracy of brothers is able to sustain itself.

Less a democracy, it seems to me, than a fraternal fascism, the brothers exclude women from civic participation and authority, forcing them into positions of "gracious submission" to men's sexual and emotional dictates.

In these all-male blood rituals of sacrifice, appeasement and empowerment was the genesis, Freud suggested, of all human ritual. Human culture is itself a complex derivation of this primal patricide, and, specifically, the taboos against patricide and incest, and not only the taboos broken by Oedipus (Greene, 1990). DiPiero (2002, 25) observes that Freud's description of the Oedipus complex—that young men "long" to murder their fathers in order to enjoy "exclusive possession" of their mothers—constitutes, in fact, a "symptom" rather than a "fundamental" and "deep structure" of manhood. Does racialization accompany such homosociality, itself inseparable from homicide, worship, and possession? Does racial phobia derive from repressed fraternal desire (see Marriott 2000, 91 n. 4)?

In Paul Hoch's (1979, 95) commentary on the primal scene, Freud, drawing upon the myths of various Western peoples, postulated that human history began with a "primal family" headed by a "repressive primal father" who maintained exclusive sexual control over the mother and their daughters, and who killed, castrated, or expelled any of the sons who challenged his authority. Eventually, Hoch notes, the sons "revolted, killed—and possibly castrated—their father, and took their pleasure with their mothers and sisters" (1979, 95). Horrified at what they had done, the sons honored their father's spirit by making of it a totem symbol and by renouncing all rights to their female kin. In this version of the primal scene we can see the genesis of the "obsessive guilt and compulsive anxiety that wrack white psychology" (Castronovo 2001, 162) and the embodied configurations of these in what Hoch (1979, 95) termed "a sexually repressed form of masculinity." It is white masculinity.

DiPiero (2002, 212) interprets Freud's myth as an account of the "ambivalent admixture" of "fear," "love," and "envy" that informs the son's relationship to the figure of the father. This Eucharistic event—after murdering the father, they eat him—constitutes the origin of "social organization, moral restrictions and of religion" (Freud 1950, 142). The principal point of Freud's story, DiPiero suggests, is to explain the power the idealized image of the father exerts in most cultures. In *The Ego and the Id*, DiPiero reminds us, Freud depicts the ego ideal as a "substitute for a longing for the father" and consequently the "germ from which all religions have evolved" (Freud 1960, 27; quoted in DiPiero 2002, 212). In the context of Genesis 9:23, it is impossible not to interpret this "longing" in sexual as well as psychological and political terms.

DiPiero suggests that the concern of *Totem and Taboo* is primarily political. What props up such patriarchal authority, Freud suggested, are guilt and fear, in the context of Noah, patriarchal guilt over Ham's (and/or Noah's) longing and the fear created by the subsequent curse. DiPiero notes that the brothers' murder of the primal father enabled them to identify with him, due to their love for him and out of their fear of each other: "Though the brothers had banded together in order to overcome their father, they were all one another's rivals in regard to the women" (Freud 1950, 144; quoted in DiPiero 2002, 212). In Genesis 9:23, there is no report of rivalry among Ham, Shem, and Japheth, no mention of "women," and the subsequent solidarity of Shem and Japheth seems, at least in part, a consequence of their father's cursing of Ham. DiPiero points to the slippage between the "ideal" and the "real," between "identification" and "desire." He suggests that the myth construes men as "resolutely hysterical," precisely because no son can ever coincide with the idealized figure of the father; more important, the myth "depends" on men's "cooperation" in the oppression of women, the very problem the myth presumably explains (DiPiero 2002, 213). After Sedgwick (1985), we recognize that such "cooperation" sublimates but does not banish same-sex desire, restructuring it as racialized homoerotic narcissistic exhibitionism.

For Freud, narcissists "are plainly seeking *themselves* as a love-object," a condition Freud discovered in those whose "libidinal development has suffered some disturbance, such as perverts and homosexuals" (Freud 1957 [1914], 88; quoted in Stokes 2001, 72). Although narcissism has come to signify significant cultural and historical ideas (see, for instance, Lasch 1978, 1984), Mason Stokes (2001, 72) understands it as a "way of being that requires a consideration of others solely for the larger purpose of articulating and buttressing the ego, the self." Freud focuses most of his attention on perversion and homosexuality; Stokes adds whiteness to these.

Freud was hardly unmindful of the association of race with incest, as David Eng reminds. Eng (2001, 8) argues that Freud "hypersexualized the primitive, racialized body." By linking sexually voracious primitives with the failure of the incest taboo, Eng (2001, 8) continues, Freud reiterated the "inseparability of racial from sexual identity." By emphasizing the "dark origins" of "primitives," Eng (2001, 8) declares, Freud "clearly connects the savage tribes with a type of visual darkness—with a type of visual marking, that of being dark-skinned." As in Genesis 9:23, the signs of racialization are not in any system of "visual authentication" (Eng 2001, 9). They are "established through Freud's depiction of the sexual practices and pathologies of primitive peoples" (Eng 2001, 9).

Because the brothers' sexuality revolves around—and is regulated by—a powerful male figure, DiPiero (2002, 214) argues that the brothers are "already united" in a "relationship" that constitutes the "cause," not the "result," of their patricide. In this regard, the sons' insurrection (and the father's resurrection) "had to have always already happened for it ever to happen in the first place" (2002, 214). Murder is the violent interruption of the incest that has already happened, if not in behavioral terms, in emotional fact. Is patricide, then, a sexualized aggression in homosocial drag?

What is the patriarch, after all? Is he (only) a convoluted and compensatory denial of lack, created in the name of the Father, expelling that "lack" as Eve? As Kaja Silverman (1988, 14) appreciates, the "equation of woman with lack [i]s a secondary construction, one which covers over earlier sacrifices." Is what is sacrificed the desire of the son for the father? Is, then, patriarchy always unstable, compensatory, hysterical? Referring to the mythology of *Totem and Taboo*, DiPiero (see 2002, 185) suggests that the (unconscious) hysteria of patriarchy becomes specularized in the obsession with anatomy, discovering in others' bodies a "physical" and "psychic lack" with which they disidentify.

Is slavery, in this sense, the displacement of that subjective bondage European sons felt after their abandonment by God-the-Father, as their identified ideal wails on the cross: "My God, My God, why have you forsaken Me?" (Mathew 27:46)? Frederick Douglass was clear that "slavery does away with fathers" (quoted in Edelman 1994b, 48), denying Africans and African Americans access to the symbolic meaning of patriarchy. It follows, does it not, that black sons would search for their lost fathers (see Marriott 2000, xiii)? Is racialized enslavement the collectivized and convoluted extroversion of the European son's sodomitical subjectivity? Is what is "black" what Europeans imagined as that "abomination" that is sexual desire among men? Is Daniel Paul Schreber the European discoverer of the New (after) World in which father and son are reunited, the second coming in which the rib is restored, enabling, finally, "*les femmes exister*"?

Psychoanalysis shares with modern philosophy, literary theory, and criticism, Ned Lukacher (1986) argues, a preoccupation with the question of origin. Psychoanalysis is dedicated to the labor of remembering the "primordial forgetfulness that conceals the origin" (1986, 26). The notion of the primal scene is key to this labor. Freud formulated the idea while working with his most famous patient, a Russian man named Sergei Pankejev. On the eve of his fourth birthday, Pankejev had dreamed that through an opened window he saw a barren tree in winter in which six or seven white wolves were sitting and staring at him, obviously about to leap onto him and consume him. He awoke screaming. For the remainder of his long life, Pankejev—named by Freud the "Wolf-Man"—never forgot the terror and the profound impression of reality that the dream created (Lukacher 1986).

In his study of the Wolf-Man's case—*From The History of an Infantile Neurosis*, published in 1918, wherein for the first time appears the concept of "primal scene"—Freud theorizes the relation of the dream to reality. Pankejev had presented Freud with both a verbal text and a line drawing of wolves sitting in a tree after remembering the dream early in the course of a four-year analysis. Much of the remaining analysis was devoted to determining the relation of the dream to reality. For nearly forty years Freud pondered the relation of dreams to reality, without ever reaching a definitive theorization. Does the dream point to the empirical fact of the primal scene, or is it the consequence of a "primal phantasy"? (Recall the controversy surrounding Freud's famous inversion of his theory that many children had been sexually molested by their parents to the theory of infantile sexuality, in which infants are themselves sexual and desire their parents.) The dream suggests something anterior, perhaps something we might characterize as "the origin," but its interpretation does not necessarily bring this "primal scene" into memory (Lukacher 1986). Can the adult son remember his submission to the Father, even when it is marked, indeed memorialized, on his "private part"?

MY BLOOD BRIDE-GROOM

The dominance of Christianity in Europe meant that Monotheistic culture would locate circumcision within a shared historical framework: a Judeo-Christian historical tradition reaching back to Abraham and, ultimately, to the story of creation in the Book of Genesis. (David L. Gollaher 2000, 44).

Castration has in fact been deeply entangled in the central beliefs and practices of Christianity for two thousand years. (Gary Taylor 2002, 14)

Circumcision . . . is ideally an injury inflicted by the father on the sons to signify their submission to God. (Howard Eilberg-Schwartz 1994, 157)

Circumcision, David L. Gollaher (2000, xi) tells us, is the "oldest enigma" in the history of surgery. He regards it as "far easier" to decode the motives for Neolithic cave painting than to understand "what inspired the ancients to cut their genitals or the genitals of their young" (2000, xi). Yet several millennia ago, long before medicine and religion were distinguishable forms of human understanding, "cutting" the foreskin of the penis was "invented" as a "symbolic wound," becoming a ritual of "extraordinary power" (Gollaher 2000, xi).

Does the practice of cutting the penis, creating a literal as well as "symbolic wound" (Gollaher 2000, 53), imply that the violation of masculinity is pleasurable? (see Silverman 1992) Is the "self-shattering ecstasy" that Leo Bersani (1995) associates with men "becoming women" a presublimated form of the sacred? Is the very structure of desire itself fetishistic, as Elizabeth Grosz (1994, 283) suggests, both affirming and denying a "founding primal object of desire while creating a substitute for it"? In this paraphrase of de Lauretis, Grosz inadvertently provides us with one meaning of circumcision: the cut penis as fetish, marking both the father's sexual desire for the son and his creation of a substitute mark for it, trading sexual possession for the genealogical kind, rationalizing the exchange religiously and medically.

The secularization of this religious practice is just over one hundred years old, Gollaher reports. It "swept" (2000, xi) America and Britain around the turn of the twentieth century. For most of the twentieth century, circumcision was the "most frequently performed" surgical procedure in the United States (Gollaher 2000, xiii). Although contemporary physicians remain divided about the risks, benefits, and ethics of the procedure, the circumcision of infants is so common that most parents and physicians scarcely think of it as surgery. This remains the case despite the fact that the American Academy of Pediatrics' Task Force on Circumcision judged in 1999 that "these [scientific] data are not sufficient to recommend routine neonatal circumcision" (quoted in Gollaher 2000, xiii). The practice is not based in science, Gollaher (2000, xiii) observes, but in "something else: tradition, experience, ritual." What are these but the traces of totem and taboo?

Although it is historically accurate to say that ancient Israel inherited circumcision from Egypt, Gollaher observes, the statement oversimplifies the complicated relation between two cultures. By the thirteenth century B.C.E., the age of Ramses II, circumcision had been established in Egypt for thousands of years. The practice was very familiar to Moses, who led the Israelites' flight from Egypt and who was influential in formulating the main elements of law, ritual, and administrative authority constituent of the Jewish nation. It was, Gollaher (2000, 6) continues, within this religious paradigm that circumcision emerged as the "characteristic mark" of Judaism.

For all his authority and influence, Moses remains an "almost ungraspable figure" (Gollaher 2000, 6). Genealogically he is described as an Israelite (Exodus 2), although he was adopted as an infant and raised among Egyptian royalty. The complexity of Moses' affiliation with the ancient Israelites has persuaded some—Gollaher names Freud—to conclude that he was in fact an Egyptian who decided to make the enslaved Hebrews' cause as his own (Gollaher 2000). Moses' reasons may not have been his own.

It is, presumably, God who commands Moses to lead the Israelites out of captivity in Egypt and to reestablish the religion of Abraham and the patriarchs.[1] It was a religion, Gollaher (2000, 7) remarks, the "defining ritual" of which was circumcision. He notes that Moses himself was not circumcised during his residence in Pharaoh's household, nor would he submit to circumcision anytime during his long life. The oldest mention of circumcision in the Torah (not in biblical chronology, Gollaher notes, but in terms of the antiquity of the source of the passage) is what Gollaher (2000, 7) terms a "cryptic" account of a confrontation among Moses, God (Yahweh), and Zipporah, Moses' Midianite wife:

> Then it happened at a stopped place along the way that Yahweh met [Moses] and tried to kill him. Then Zipporah took a piece of flint and cut off her son's foreskin and touched [Moses'] feet with it, saying, "You are my blood bride-groom." So [Yahweh] let him alone. At that time she said "blood bride-groom" in reference to circumcision. (Exodus 4:24–26)

In ancient Judaism, blood, not semen, seems the medium of the son's submission.

The sources of this passage are, Gollaher (2000, 7) tells us, "extremely obscure." He notes that it provoked endless disputation among Jewish and Christian scholars who labored to reconcile it with those portraits of Moses and his relationship to God as presented in other parts of the Bible. Among the main points of the passage are: (a) the baby's circumcision by his mother (b) the mother's touching of the father's genitals ("feet") with the son's severed foreskin (a sexual act between two males mediated by the female body); and (c) the "magical transference" of circumcision from Moses' infant son to the (uncircumcised) father (Gollaher 2000, 7). Gollaher suggests that the phrase "blood bride-groom" may refer to an earlier time in Israel's history when circumcision may have been a premarriage initiation ritual, preparing the bridegroom for heterosexual coitus. But in the Exodus passage quoted above, the preparation is for "marriage" between God and Moses.

Although a "cornerstone" of Judaism, Gollaher (2000, 7) notes, circumcision fits into the biblical narrative in a peculiar fashion. He points out that Moses delivers a divine law with which he then fails to comply. But Moses' uncircumcised state is only one of several peculiarities. Despite his heroic compliance with God's command to lead the Israelites out of Egypt, Moses himself was prevented from entering the promised land. Moreover, unlike the patriarchs, Moses was denied the honor of being buried there. Several rabbinic commentators suggest that this was God's punishment of Moses for not having been circumcised. Rabbi Joshua Ben Karha declared in the *Mishna-Nedarim*: "Great is the precept of circumcision for neglect of which Moses did not have his punishment suspended for even a single hour" (3:11). Gollaher quotes Old Testament scholar Peter Machinist who suggests that Moses, in his "strangeness," amounts to a kind of antihero, "someone who does not serve the native tradition at any point as a role model who can really be emulated" (quoted passages in Gollaher 2000, 8).

Is there a relationship between Moses' leading the Israelites out of captivity and the enslaving curse of Ham? Is Moses Ham almost redeemed, the condemned son who expiates his otherness by leading the future generations out of the enslavement his father's rage has required, to another form of enslavement now disguised as devout devotion to God? Is circumcision, then, a sign of sublimated enslavement, of the son's chosen subjugation to an abstract, that is, disembodied father? Is the cut penis a (t)race of his desire for the father, desire that is the cause of both his enslavement and its effect, spiritual "salvation"?

It is Abraham, not Moses, whom the Torah makes as a model of God's son, Gollaher points out. It is to Abraham that God discloses himself; it is Abraham's children who become God's chosen people. Early on, the ancient Israelites came to believe that God had promised Abraham that he will make of the Jews a great nation. But this promise was conditional, Gollaher notes: it depended on Abraham's obedience, his observance of the covenant between them. According to the Genesis narrative, at the center of this covenant was circumcision, presumably an outward symbol of Abraham's good faith, his obedience, indeed, his subjugation to God-the-Father:

> God said to Abraham, "For your part, you must keep my covenant, you and your descendants after you, generation by generation. This is how you shall keep my covenant between myself and you and your descendants after you: circumcise yourselves, every male among you. You shall circumcise the flesh of your foreskin, and it shall be the sign of the covenant between us. Every male among you in every generation shall be circumcised on the eighth day, both those born in your house and any foreigner, not of your blood but bought with your money. Circumcise both those born in your house and those bought with your money; thus shall my covenant be marked in your flesh as an everlasting covenant. (Genesis 17:10–13)

Abraham circumcised his son Ishmael at age thirteen. The mark of the Father is branded into the flesh of his sons and his sons' slaves, a sign of subjugation and possession through the generations.

Any idea that strong parallels exist between a primary ritual of Judaism and those brutal and bloody rites of passage observed in "primitive" societies (such as the Sambia) seems offensive, even sacrilegious, to many (Gollaher 2000). Even if the cutting of the foreskin has often been practiced as a fertility ritual, observed the historian Roland DeVaux, Israel's monotheistic "religion gave the ritual a more lofty significance" (quoted in Gollaher 2000, 9). Are Western religions, in their lofty patriarchal pretensions of transcendence, deferred displacements of homoerotic desire generally and of father–son desire specifically? Is that why women have, historically, been so marginalized within these religions?

Gollaher recalls Eilberg-Schwartz's point that the Hebrew word characterizing the relationship between covenant and circumcision makes clear that the two are integrally related; Gollaher (2000, 9) observes: "Circumcision, in other words, was not merely a sign of the covenant; it constituted a vital part of the promise itself. In a sense circumcision *was* the covenant." A covenant, Gollaher (2000, 9) reminds us, is a "sacred agreement." Is the agreement that neither father nor son will confess that each is emasculated and mutilated, each bound to patriarchy? This sleight of hand is achieved through sublimation, that is, reproduction, the father's, in Abraham's case God's, promise to grant him a miraculous fertility. God told Abraham:

> This is my covenant with you: You shall be the father of a multitude of nations. And you shall not longer be called Abram, but your name shall be Abraham, for I make you the father of a multitude of nations. I will make you exceedingly fertile, and make nations of you; and kings shall come forth from you. (Genesis 17:4–6).

This promise (is it not the consolation prize?) to Abraham included a threat: "Every uncircumcised male, everyone who has not had the flesh of his foreskin circumcised, shall be cut off from his people. He has broken my covenant" (Genesis 17:14). In other words, the uncircumcised are to be expelled from the community. Disobey the Father and ye shall become Other.

Gollaher reminds us that among a desert-dwelling tribe such exile amounted to a death sentence. Some medieval commentators suggested that "cut off from my people" may also have meant that the uncircumcised would suffer the curse of infertility or impotence. In a patriarchal culture this was almost as serious a threat as expulsion, as the inability to reproduce was "bitterly disgraceful" (Gollaher 2000, 10). The son who declines to accept identification in place of desire, and who declines to reproduce, is cursed. The "degenerate" is the son who declines to accept the scar in place of the sexual, who remains loyal to the father by declining to *become* him. This son discerns the "other" within the "same."

We are told that Moses suffered a speech impediment: on two occasions he is described in the Torah as suffering from uncircumcised lips. More generally, Gollaher continues, the term *uncircumcised* was used to slur the Philistines (1 Samuel 18:25), suggesting that because they were excluded from the covenant with God, they constituted a lower order of being. Is this the origin of the Great Chain of Being? Its use as a slur rather than a descriptor is consistent with many biblical writers' use of "circumcision" as a metaphor, not a physical fact. To illustrate, Gollaher points to the following characteristic passage in Deuteronomy, wherein the one who resists God is admonished to "circumcise therefore the foreskin of your heart, and be no more stiff-necked" (Deuteronomy 10:16).

In other passages, the circumcision of the heart is depicted as a divine act, a kind of spiritual surgical procedure. "And the Lord your God will circumcise your heart, and the heart of your offspring, to love the Lord your God with all your heart [love your Master with all your heart], and with all your soul, that you may live" (Deuteronomy 30:6). Gollaher notes that the prophet Jeremiah employs the same phrase in order to distinguish between nominal circumcision (of the foreskin) and true circumcision (of the heart). What can it mean to circumcise the heart? To accept positions of "gracious submission," as Southern baptists depicted wives' relation to their husbands? "The time will come when I will punish all the circumcised that are uncircumcised," Jeremiah warns unbelievers. "For all the nations are uncircumcised, but Israel is uncircumcised at heart" (Jeremiah 9:25–26). He adds that those who do not obey God's words suffer uncircumcised ears (Gollaher 2000). There are apertures yet to be filled.

From embodied desire the ancient Israelites abstracted the cultural practices they were sure God-the-Father demanded. Gollaher notes that among the dietary laws listed in Leviticus we find the following passage:

> When you enter the land and plant any tree for food, you shall regard its fruit as its foreskin. Three years it shall be uncircumcised for you, not be eaten. In the fourth year all its fruit shall be set aside for jubilation before the Lord; and only in the fifth year may you use its fruit—that its yield to you may be increased: I am the Lord your God. (Leviticus 19:23–25)

Gollaher explains that fruit trees growing in Israel, among them figs, olives, grapes, and dates, typically produce little fruit during their early years. Their capacity to "bear fruit" comes later, as the writer acknowledges. In this sense, the trees are likened to the uncircumcised boy, whose potency awaits the removal of his foreskin in preparation for heterosexual intercourse and paternity (Gollaher 2000).

Is this not all intolerably queer? The tell-tale sign is the centrality of circumcision to the covenant; it points to the "forgotten" trauma between father and son for which the covenant substitutes. In turn, the covenant (re)produces trauma, and not only between father and son. Moreover, as a fetish, circumcision enables the son to not only elude paternal prohibition (in fact he is forced to accede to it), it also provides a means by which he can continue undisturbed—and out in plain sight—in his gratifying sexual activities, including the homoerotic enjoyment of other men (see Eng 2001, 146). Among the Sambia, the covenant between son to the father is, well, not so sublimated.

COMING OF AGE

Circumcision . . . beautifies men in God's eyes. (Howard Eilberg-Schwartz 1994, 171)

There is no necessary reason for identification to oppose desire, or for desire to be fueled by repudiation. (Judith Butler 1997, 149)

Identification and desire are complexly imbricated with each other—so much so that it is often possible to uncover the former through the latter. (Kaja Silverman 1988, 216)

What does it mean—what difference does it make—when a social or political relationship is sexualized? (Eve Kosofsky Sedgwick 1985, 5)

The Sambia of New Guinea have been studied extensively, most famously by anthropologist Gilbert Herdt, whose research has been reviewed by, among others, David Gilmore. The Sambia, Gilmore (1990, 146) explains, are of "unusual interest" to anthropologists due to the "intensity" and, to

"Monotheistic eyes," the "perversity" of their masculine rites of passage. I might add "straight" monotheistic eyes, although no doubt there will be queer readers who also find these rites strange.

I am referring, of course, to the ritualized practices of homosexual fellatio, bloodletting, and hazing. It is this "oddity" of Sambia rites of masculine passage, and specifically the ritualization of homosexuality, that earns the Sambia the status, for Gilmore (1990, 146), of "an important test case for our study of manhood images." "How does," he asks, "this homosexual passage fit in with the hyper-heterosexuality that we have seen before in manhood codes?" (Gilmore 1990, 146–147).

The Sambia are a people obsessed with masculinity, Gilmore reports. They regard masculinity as "highly problematic," indeed, a "quandary" and a "penance" (Gilmore 1990, 147). (Already they seem a wise people.) Gilmore writes that "they"—does he mean everyone, or only "men"—are "firmly convinced" that "manhood" is artificial; it must be "forcibly" "induced" by "ritual means" (1990, 147). The Sambia require their young men to endure a painful process of induction into manhood through sequenced rites of transition (Herdt 1981, 1982). Suddenly, it seems, Gilmore (1990, 147, italics added) gets to the point: "What makes the Sambia special, even unique, is their phase of ritual homosexuality in which youngsters are forced to perform fellatio on grown men, *not for pleasure*, but in order to ingest their semen. This then supposedly provides them with the substance or 'seed' of a growing masculinity." How does Gilmore know pleasure is not involved?

To illustrate what is at stake in this ritualized fellatio, Gilmore quotes one of Herdt's informants, Tali, a Sambia ritual expert: "If a boy doesn't 'eat' semen, he remains small and weak" (Herdt 1981, 1; quoted in 1990, 147). Gilmore quickly adds that this "homosexual phase" is "only temporary" (Gilmore 1990, 147). Fellatio is followed by an adult life of "full heterosexuality" (Herdt 1981, 3; quoted in Gilmore 1990, 147), including marriage, procreation, and "all the more usual masculine virtues" (Gilmore 1990, 147). The ritualization of homosexuality, therefore, is a passage to "masculinization" (Herdt 1981, 205; quoted in Gilmore 1990, 147), and it is this apparently "contradictory" relationship between "means" and "ends" that renders the Sambia "interesting" and "important" (Gilmore 1990, 147).

Sambian conceptions of manhood derive from mythic depictions of their past lives as warriors and their present ones as hunters: they picture masculinity as aggression based on courage and stamina. Sambia men, then, espouse "a particular conception of manhood" (Herdt 1981, 16; quoted in Gilmore 1990, 150) extolling toughness, indifference to danger and pain, decisive action, physical strength, and risk-taking. The embodiment of this ideal is the warrior-leader, a man who serves as a model to all other men. If "model" means a man with whom other men can identify, then the homoerotic undertow must be considerable (Butler 1997). Is that why the Sambian conception of manhood is exactly contrary to their conception of women?

Sambian ritualized homosexuality is hardly unprecedented in human history, as Herdt recognized. Herdt discussed the parallels between the Sambia and the ancient Spartans, who, like other Greeks of the time, also practiced a "manly" homosexuality, including on the battlefield (Vanggaard 1972; Gilmore 1990; Herdt 1981). In each instance, the warrior ethic not only permitted but also encouraged sexual relations between soldiers and boys. These liaisons were not coded as effeminate but as masculine and as intensifying warrior resolve and solidarity (Gilmore 1990; Halperin 1990).

Gilmore points out that in both cases (in Sparta and Sambialand), the soldier took, presumably, the active role in the sex act, that is, that of the penetrator. The soldier's lover or partner was often a youth of inferior status who acted out the passive "feminine" role, although David Halperin's (1990) research on ancient Athens would seem to suggest a more complicated, if still hierarchical, sexual configuration. Gilmore's information suggests that the ancient Greek lover was usually a youthful slave who was sodomized by an adult in a superior status (Veyne 1958), a view Halperin's research contests. Both agree that lovers in both instances were young; in New Guinea Herdt studied prepubescent boys, not yet "men." In this sense, Gilmore suggests, the sex involved is, from a Sambian perspective, not precisely homosexual, if by that term there is some suggestion of consenting adults.

Among the Sambia, Gilmore tells us, adult homosexual relations do not seem to exist. Due to this fact and due to the ritualized and transient character of man–boy relationships, "homosexual" may not be, Gilmore asserts, the most appropriate term to employ in characterizing Sambia practices. "As the fellatio is a means to an end rather than an end in itself," he continues, "ritualized masculinization may be a more accurate (and less ethnocentric) term" (Gilmore 1990, 151). Given that fellatio in this culture is also not always "an end in itself," but provides a passage to masculinity, perhaps conservative politicians will want to rethink their conceptualization of (and opposition to) it?

Interestingly in terms of object-relations theory and Laqueur's (1990) one-sex theory, the Sambia seem to have a single-sex theory of gender development. It is not masculine. Femininity is thought to occur naturally, an internal maturation in continuous association with the mother. In contrast, masculinity is no inevitable result of anatomical maleness; "it is an achievement distinct from the mere endowment of male genitals" (Herdt 1982, 54; quoted in Gilmore 1990, 152). To put the matter another way, while girls presumably become women "naturally" because they retain ties to their mothers, boys have to be made into men by other means. In other words, "masculinity must be achieved" (Herdt 1982, 55; quoted in Gilmore 1990, 152). Gilmore (1990, 152) quips: "All this makes one wonder if the Sambia have been reading the neo-Freudians."

This single-sex view—Gilmore (1990, 152) wants to characterize it as "dualistic" which, I agree, it becomes—of sexual maturation derives from Sambian notions of physiology. The Sambia believe that women are born with an internal organ called the *tingu*, an organ responsible for their "natural" evolution into women. Sons are born with the same organ, but they are constitutionally inferior in this regard: their *tingu* is weak and inactive and requires semen to grow (Herdt 1981; Gilmore 1990). This would seem to be the castration tale told in reverse. Born of women, sons must be made into men, a developmental accomplishment that occurs only through ritualized "insemination," which triggers and sustains the process, and through the strict and careful guidance by male elders. "The key to all this," as Gilmore (1990, 152) sees it, "is to get the boy away from the baneful influence of his mother so that this *tingu* may be stimulated to grow and implement the masculinization process. Closeness to the mother prevents this, pulling the boy back to a sexually indeterminate infantilism." This view is hardly limited to the Sambia (see Pinar 2001, chapter 6).

Though "extreme" for the New Guinea Highlands, Gilmore (1990, 152) tells us, Sambian beliefs about gender and sexual maturation are not unique. In fact, such ideas are rather common in the Highlands; the Sambia demonstrate an "extreme version" of a "widespread" and "passionate" belief in the "artificiality" of manhood (Gilmore 1990, 153). Among the Etoro, Onabasulu, and Kaluli tribes, for example, there is a shared belief that "boys do not become physically mature as a result of natural processes" (Kelly 1974, 16; quoted in Gilmore 1990, 153). Like the Sambia, these peoples believe that the "growth and attainment of physiological maturation is contingent upon the cultural process of initiation" (Kelly 1974, 16; Schieffelin 1982, 162; quoted in Gilmore 1990, 153). Likewise, among the nearby Gururumba tribe, sons do not mature into men naturally but must be "made" into men (Newman and Boyd 1982; Gilmore 1990).

In each of these cases, the rituals of maturation into manhood include what in the West we would characterize as homosexual fellatio. There are Highland tribes, however, the Mountain Ok, who share the same basic beliefs about the fragility of manhood but without ritualized homosexuality (Barth 1987; Gilmore 1990). Other tribes have held firm in many of their beliefs but have abandoned the fellatio ritual due to the outrage of white Australian administrators (Gilmore 1990). For all, Gilmore (1990, 153) reports, these convictions are "intensely held" and cause "unrelenting anxiety." Why? Comparative studies make clear the universality among Highland peoples of this belief in the fragility of masculinity. Gilmore quotes Roger Keesing who, in his comprehensive survey of New Guinea male cults, concluded that their most distinctive feature is an emphasis on male gender as a created "rather than a natural" consequence of maturation (1982, 5; quoted in Gilmore 1990, 153).

Appreciating this point goes a long way, Gilmore suggests, in explaining the cultural meaning of ritual fellatio in Sambia male initiation rites. Young men—boys—need to ingest semen in order to masculinize their bodies. The focus upon semen as the key marker of masculinity is not completely different from the emphasis, in American culture, upon the erect penis as a prerequisite to fully functioning (phallocentric) manhood. (Impotence is, for such phallocentric men, a crisis in gender identity: witness the popularity of Viagra and other erection-producing drugs.) Gilmore, too, sees the throughlines among cultures, noting that, aside from the use of fellatio as the method of masculinization, the notion that men are "made"—not born—should come as no great surprise. "There are," he continues, "underlying similarities in motivation across cultural boundaries, if not in the actual practice" (Gilmore 1990, 154).

Gilmore calls upon the ancient Greeks (citing Dover 1978) to point out that (using for the moment nineteenth-century language) "sexual inversion per se is not universally linked to a lack of masculinity in all the major Monotheistic traditions, nor is pederasty always a sign of effeminacy" (Gilmore 1990, 154). (Tell that to the prosecutors of Michael Jackson.) During much of the ancient Greek and Roman periods "lovers of boys were just as numerous as lovers of women" in classical antiquity (Veyne 1985, 28; quoted in Gilmore 1990, 154). Gilmore (1990, 154) calls these men "Greek and Roman homosexuals" (despite the consensus, after Foucault, that such a sexual identity is a recent twentieth-century phenomenon; see, again, Halperin 1990), noting that these men, in loving boys, did not relinquish their manhood. Ancient Greeks and Romans remained "men" as long as they expressed their desire in "active" positions, that is, as penetrators (Gilmore 1990). That hypersexuality was considered an African—specifically, an Ethiopian—characteristic expressed, David Brakke (2001, 513) suggests, "Roman anxieties about legitimacy and power," not racialism (see Brakke 2001, 511).

Gilmore quotes Kenneth Dover's study of homosexuality in ancient Greece; Dover (1978, 106; quoted in Gilmore 1990, 154) concluded that the "abandonment of masculinity" occurred only if an adult man accepted the passive or receptive role in the sex act. Such a role meant relinquishing manly control and dominance, a point congruent with Halperin's (1990) research. Evidently the same was true of the ancient Romans, who believed that "to be active was to be male, whatever the sex of the compliant partner" (Veyne 1985, 29; quoted in Gilmore 1990, 154–155). In his *Dialogues on Love*, Plutarch repeats the prejudice: "Those who enjoy playing the passive role we treat as the lowest of the low, and we have not the slightest degree of respect or affection for them" (quoted in Gilmore 1990, 155). Gilmore tells us that this preference of position is found in modern Greek culture as well, citing Peter Loizos' finding (1975) that many Cypriot Greeks distinguish between the *poushtis*, a man who takes a "passive" or womanlike role in sex, and his homosexual active partner. The poushtis, or *poustis*, as he is called on the mainland (Campbell 1964), is strongly denigrated, but the "active" participant retains his manhood, precisely because he takes the "manly" position of penetrator (Gilmore 1990, 154). This appears to be the case as well in contemporary Latin America (see Almaguer, quoted in Bordo 1994, 289).

In ancient Sparta, lovers were assumed to be more intense warriors, precisely *because* they had their lovers by their side on the battlefield. (Was this the case, one wonders, even when one soldier was sexually "passive"?) But the ancient Greeks were hardly alone in their preferences. So-called transient homosexuality (Gilmore 1990, 155) was common in the preindustrial warrior societies throughout monotheistic Europe, including northern

Europe. Gilmore recalls Dover's report that, in medieval Scandinavia, for instance, men could be "men" even with male lovers, as long as they kept the "active" role in the encounter: "In the old Norse epics the allegation 'X uses Y as his wife' is an intolerable insult to Y but casts no adverse reflection on the morals of X" (Dover 1978, 105; quoted in Gilmore 1990, 155).

Much the same was the case among the Japanese samurai. Gilmore recalls Ian Buruma's (1984) finding that for many centuries homosexuality was not only tolerated in Japan but was, in fact, encouraged as a purer form of love. In Japan, "as in Sparta or Prussia . . . gay lovers make good soldiers, or so it was hoped" (Buruma 1984, 128; quoted in Gilmore 1990, 155). Gilmore also notes that German militarists in the 1920s and 1930s held similar ideas, including members of the proto-fascist Freikorps, as Klaus Theweleit theorizes in his *Male Fantasies* (1987). George Mosse (1996) situates German forms of masculinism historically and within European culture, as this book explores.

Returning to the New Guinea case (not as unique, we now see, as at first it might have seemed), Gilmore (1990) argues that what distinguishes the Sambia from other warrior civilizations is the formal ritualization of the passage to manhood, not its erotic content. He points out that the Sambia ritual transition occurs over several years and includes numerous ceremonies in addition to ritualized fellatio. The process begins with what Gilmore (1990, 156) terms "the most important single event in the male life cycle: the physical separation of boys from their mothers." This is, he continues, a "dramatic rupture" (1990, 156) that constitutes the first step in a male-conceived, male-dominated process of "masculinization." The elder men build "an all-male cult house" (Gilmore 1990, 156) where the young men spend much of their time as initiates. Abducted from their mothers, now in the company of older men, Sambian boys are forced to follow numerous symbolic and psychological rites of transition and induction (Gilmore 1990).

Common to other male initiation rituals in other parts of New Guinea, Sambian boys are moved where their mothers cannot see them (and where they cannot see their mothers), often to an all-male place in the bush. There they are subjected to "brutal hazing" (Gilmore 1990, 56), involving both physical beating and/or painful bloodletting. One of these practices is nose-bleeding, in which instance the boys are forced to make blood flow from their nostrils. In the past the initiates were made to force stout bamboo canes down the esophagus, causing both bleeding and painful vomiting, but this practice, Gilmore tells us, has been abandoned. Herdt (1981) describes the nose-bleeding as the single most painful ritual act and constitutes, whatever one's cultural location, physical and psychological trauma. Stiff, sharp grasses are thrust up the boys' nostrils until blood flows copiously, a practice analogous to rape, it seems to me.

The nostril is the aperture, the stick of grass the phallus; the boys are rendered menstruating "girls." Having brought the boys into submission by this invasive procedure, the older men respond to the flow of blood with a collective war cry (Herdt 1981; Gilmore 1990). "As they will have to do the battlefield later," Gilmore (1990, 156) concludes, desexualizing the assault, "the boys have shown fortitude and have learned to disdain the shedding of their own blood." Like sons marked by a circumcised penis, Sambian boys will not forget their covenant with the fathers.

Now that that the young men have been bleeding, both literally and figuratively, they are easy prey for various other rites of both ingestion and egestion and by ritual flogging with ceremonial objects. The boys are now beaten violently with sticks, switches, or bristly objects until their skin is "opened up" and once again blood flows (Herdt 1981; Gilmore 1990). No efforts are made to mitigate the initiates' terror or pain; in fact, Gilmore (1990, 156) suggests (once again desexualizing the practice), "overcoming such agonies seems to be the point." On this point he is relying on Herdt, who writes that these practices teach the young men to ignore the flow of their own blood and to show a stoic resolve, preparing them for the life of "manly" endurance expected of them. They are also taught to subtract the sex in violent male–male sexual sadism.

Such practices take place in other parts of the Highlands, Gilmore tells us, relying on a range of research (cf. Read 1965; Newman and Boyd 1982; Keesing 1982), where boys are presumably also toughened up by physical beatings. In particular, Gilmore cites Fitz John Poole's (1982) reports of violent male rites among the Bimin-Kuskusmin, rites that terrorize and traumatize young boys. Understandably one ethnographer was unable to retain a pose of researcher neutrality in the face of such violence, describing the male initiation practices of the Ilahita Arapesh as "cruel, brutal, and sadistic" (Tuzin 1982, 325; quoted in Gilmore 1990, 157).

The rite of homosexual fellatio constitutes, Gilmore reports, the culmination of the initiation. In contrast to nose-bleeding and beating and other forms of physical assault, this fellatio "involves no physical pain and demands *only* submission" (1990, 157, italics added). Men who have been raped orally are not likely to concur with Gilmore's characterization (see Scarce 1997). Herdt describes the Sambia as being too "prudish" to do this publicly, but this sounds like a bourgeois projection. Whatever the motives, ritualized fellatio is conducted in private between individuals, usually under the cover of darkness, leaving one to wonder how gentle or how violent the oral entries are. Gilmore (1990, 157) describes the scene: "The boys are forced repeatedly to suck on the penises and swallow the semen of the older men; the ingested semen passes down into the inactive semen organ, where it is absorbed and accumulated." Not only do the Sambia believe that repeated inseminations "create a pool of maleness" (Herdt 1981, 236; quoted in Gilmore (1990, 157), they also believe that ingested semen strengthens a young man's bones (got milk?) and builds his muscles, a view rather absent from President John F. Kennedy's physical fitness campaign (Griswold 1998; Pinar 2004a). If enough semen has been ingested by the time the boy begins puberty, facial hair appears (Gilmore 1990).

Why is the covenant between father and son made away from the mother, inside Noah's tent? Why does identification with the father require sacrifice, pain, stoicism? Why must openings be penetrated, why must the penis be branded? In Western versions, the mohel sucks the son's penis to draw the blood, to heal the wound. In the desublimated versions, the sons suck their fathers'. Is worship a sublimated form of fellatio? Does Christian prayer reinscribe the position of the Sambian supplicant?

GOD-THE-FATHER

[R]eligion is a human projection. (Howard Eilberg-Schwartz 1994, 14)

The perfect disciplinary apparatus would make it possible for single gaze to see everything constantly. (Michel Foucault 1995, 173).

Castration is always counteracted by prosthetic replacement. (Thomas DiPiero 2002, 176)

[T]he white soul is the prison of the black body. (Russ Castronovo 2001, 168)

Does the concept of a "disembodied" God derive from "discomfort" with the idea of "God's penis?" Howard Eilberg-Schwartz asks (1994, 1). God's penis could be a problem for those men commanded not to lie with men as they lie with women. The idea of an incorporeal God derives, Eilberg-Schwartz speculates, from the sexual tension men might feel in a relationship with a God who is explicitly male. Eilberg-Schwartz is interested in "fatherhood" and how the "sexual body" of a "father God" troubles masculinity (Eilberg-Schwartz 1994, 1). The penises of the fathers did not seem troubling for Sambian sons.

That the God of Jews and Christians is gendered male functions, as many feminists have underscored, to provide theological legitimation for patriarchy (see, for instance, Daly 1978). Although acknowledging this political and cultural function of a God gendered male, Eilberg-Schwartz adds that the fact destabilizes masculinity as well. In "ancient Judaism," a term Eilberg-Schwartz (1994, 2) employs to depict the various religious cultures of ancient Jews, from the period of Israelite religion under the monarch (ninth century B.C.E.) through the rabbinic period (200–600 C.E.), the maleness of God posed for men the dilemma of homoeroticism. He points out that the love of a man for his male God was sexually tense in ancient Israel because the divine–human relationship was often described in erotic and sexual terms. Marriage and sexuality are common biblical metaphors for describing God's relationship with Israel (Eilberg-Schwartz 1994; see also, Schwartz 1997).

God is imagined by these biblical (probably) male writers as the husband to Israel the wife; espousal and even sexual intercourse are metaphors for the covenant. (Consistent with this logic, then, is the religious ritual of circumcision which "castrates" the son so he may become a "bride" of God). When Israel strays from God-the-Father, "she" is judged as "whoring." Israel's relationship with God is to be a monogamous one; idolatry constituted adultery. "[S]elf-arousal," Thomas Laqueur (2003, 121) asserts, was also a form of "idolatry." Is not masturbation homosexual, at least behaviorally?

Eilberg-Schwartz notes that it was ancient Israelite men, not women, who were imagined to have the primary intimate "relations" with God-the-Father. While Israel was imagined as a woman, it was a woman who is actually a man, or men, among them Moses, Noah, and the patriarchs who loved, who served, in ways that were imagined erotically and sensually, God-the-Father.

Such a homoerotic relationship between man and God might not have posed a problem, Eilberg-Schwartz suggests, had not ancient Israelite men been so strongly pressured to procreate. Being a man in ancient Israelite culture required marriage, required fathering children, required extending the lineage of one's father and tribe. Being a man, then, was not—in sexual function at least—different from what is required now: being what, in the twentieth century, has become known as "heterosexual." In ancient Judaism, "woman" was imagined as the natural counterpart of "man," and sexual acts between men were condemned as abominations (Eilberg-Schwartz 1994). By definition, condemnations are defensive and compensatory. Those guys must have been all over each other.

The sexual politics of ancient Judaism were founded on the fantasy that "woman" was complementary to "man," and that marriage was a return to primordial unity, restoring the rib as it were. At the same time, a man's relationship to God was characterized as loving and sensual, even sexualized. It is for these reasons, Eilberg-Schwartz (1994, 3) proposes, that "various" myths and rituals of ancient Judaism functioned to "suppress" the homoeroticism implicit in the men's relationship with God. These attempts take two significant forms, he argues: (a) a prohibition against depicting God (i.e. covering the body of the Father) and (b) the feminization of men. By transgendering men as wives of God, Eilberg-Schwartz speculates, the ancient Israelites institutionalized a "heterosexualized" notion of gendered complementarity that supported procreation.

By feminizing men, Eilberg-Schwartz continues, men were rendered irrelevant, as women were imagined as the "natural" partners of a divine male, an idea evident in the Immaculate Conception story in the "new" testament. Eilberg-Schwartz suggests that this inadvertent irrelevance of men may help explain misogynist tendencies in ancient Judaism. But one must point out that by imagining themselves feminized—and, given the Genesis creation story, "woman" is imagined as a fragment of "man"—men preserved their centrality in the relationship of God-the-Father to his children, children who, given this imagery, turn out to be all sons, even when they are his "rib."

If religion is a human projection, it is specifically a *man*'s projection, an idea understood differently, Eilberg-Schwartz points out, within psychoanalytic and feminist traditions. For Freud, religion reflects and repeats the experience of having a father. At times in Freud's theorization, God symbolizes every child's (the daughter's as well as the son's) experience of having a father. At other times, it is the son's oedipal struggles, specifically his feelings of love, hate, and competition toward his father, that are projected onto a vengeful Yahweh. At still other times, Freud (1927) imagined God as providing solace and consolation in a world of pain and suffering. Finally, Freud claimed that religion is a memory of an actual historical event in which the brothers killed and ate their father (Freud 1912–1913; Eilberg-Schwartz 1994), an event studied in chapter 1, "The Primal Scene."

Does cannibalism make explicit the "oralization" of identification, that is, the aggressive incorporation of the father, not dissimilar from Sambian fellatio rituals? If so, is identification an aggression against both one's subjective self *and* the other with whom one identifies? Does identification dismember and make disappear the one identified, as he is incorporated into the body of the son who would identify with him? Would not desire be preferable to this intrapsychic violence, if it keeps (potentially, at least) separate and reciprocal the two? Is that how the spiritual father became split from the human father? The father whom the son desired was disembodied, made abstract and sacred, then worshipped. Only on the spiritual plane could the son be his father's lover; the point is, Eilberg-Schwartz (1994, 15) emphasizes, that for Freud the experience of divinity derives from and is forever "implicated" in the "experience" of having a father. Daniel Paul Schreber was transported to this spiritual plane and lived to tell about it, as we will see.

Eilberg-Schwartz argues that it is not the experience of having a father that is projected heavenward: it is a fantasy of masculinity. This fantasy is a representation of those concrete social relations in which the most valued social prerogatives—such as political power—belong to men. What it means to be a man is produced in the dynamics of the God–son relationship, including its implied master–slave dynamics, specifically the threat and reality of harsh punishment coupled with total forgiveness and (spiritual) subjugation. Do these "spiritual" dynamics construct an European culture prepared for three hundred years of sexualized domination and economic exploitation of Africans?

The disembodied notion of a God-the-Father deifies a patriarchal social order by structuring the subjectivity of "Western" masculinity. This structuring requires men's projective representation of femininity, against which they then define themselves (Eilberg-Schwartz 1994). It requires European men's projective representation of "race," against which they would fantasize their desexualized and disembodied "spiritual" natures. Feminist studies of religion, Eilberg-Schwartz (see 1994, 15), summarizes, have thus tended to employ either a correspondence or legitimation theory of male projection. Both speak to European men's racial projections as well.

Psychoanalytic (especially Freud's) theory emphasizes, Eilberg-Schwartz points out, conflict in men's religious projections. It presupposes tensions among different masculinities. If God is the experience of father symbolized theologically, then divine masculinity can be no simple confirmation of human masculinity. To underscore his point, and to emphasize a distinction between psychoanalytic and feminist theories, Eilberg-Schwartz (1994, 16) suggests that while psychoanalytic theory tends to regard religion as a "projection" of a "desirable" if "unattainable ideal," feminist theory tends to regard it as a reflection of a "problematic real." These are not mutually exclusive analyses, of course, as Eilberg-Schwartz (1994, 16) appreciates:

> Masculinity is threatened by the very constructions that seem to make it possible in the first place, and human men are diminished and challenged by the projection that authorizes their power and social position. Images of deities, of which a divine father is one primary example, thus do more than simply reflect the social order; they challenge and subvert it as well.

In our time, they seem, on balance, to threaten the social order, as Christian and Islamic fundamentalisms assault secular modernism (see Armstrong 2001).

A masculine God, Eilberg-Schwartz (1994, 17) suggests, is a "male beauty image" in the sense that it represents an ideal against which men measure themselves and in terms of which they fall short, a point Thomas DiPiero (2002) emphasizes as well. Is this a fundamental (if historical) "structure" in the monotheistic male mind that psychoculturally enabled European men to enslave the black African man they come later to fear as a phallic god? Is the slavemaster the oedipal son who demands the patriarchal position to deflect the "unmanning" desire, not only of but for his father, rendered explicit in the case of Daniel Paul Schreber? As Eilberg-Schwartz observes, Freud knew that boys not only identify and compete with their fathers, they also are, and desire to be, the objects of their father's gaze. This latter desire implies that a boy wants to be castrated, to become a "woman," which is precisely what Schreber experienced while in the gaze of God-the-Father.

Schreber was hardly the first man who experienced his love for God, and God's love for him, as feminizing. Eilberg-Schwartz (1994, 18) quotes Caroline Walker Bynum's (1982, 161) analysis of twelfth-century Cistercian images of Jesus as Mother:

> Given the twelfth-century partiality for metaphors drawn from human relationships, religious males had a problem. For if the God with whom they wished to unite was spoken of in male language, it was hard to use the metaphor of sexual union unless they saw themselves as female.

Centuries later Protestant men imagined themselves to be "brides of Christ" (Greven 1977), a sexually threatening gendering of religious faith that led to its repudiation and reformulation as "muscular" Christianity (see Pinar 2001, chapter 5, section VIII).

While an unstable repression, the son's sexual desire for the father is sublimated, then, through religious faith. So sublimated, Eilberg-Schwartz (1994, 25) notes, "God is a masculine deity whose maleness is repressed and avoided. People do think of God as a he *without* a male body." So repressed, the issue of a divine penis (or, even, a beard) seems unsettling. Despite the ancient Israelite emphasis upon the invisibility of God, Jews sometimes did imagine God in human form and later, so did Christians, fantasies portrayed on one occasion, for instance, on the ceiling of the Sistine Chapel. The absurdity of a divine phallus points to a series of tensions within masculinity as it is constructed in monotheism (Eilberg-Schwartz 1994). To unpack these tensions, Eilberg-Schwartz (1994, 26) recommends, we must face the "meaning" of the "father's nakedness," the homoeroticism "implicit" in monotheism, the compulsion to procreate, fantasies concerning conception, creativity, circumcision, and "much more." I focus here on circumcision.

At this point Eilberg-Schwartz introduces Lacan, acknowledging that Lacan's definition of the phallus has been interpreted variously: in one view it is a symbol of each child's entry into culture, in another, of each child's coming to speech. As it did for Sambian boys, becoming a human subject involves castration, the loss of the phallus. Like object-relations theorists, Lacan emphasizes the preoedipal phase of the child's development. Lacan emphasizes that human subjectivity originates in traumatic loss, a loss necessitated by the differentiation of the baby from the mother. This loss accompanies the child's entrance into the symbolic order, into culture; it precipitates desire. Human desire always seeks—and fails—to re-experience the wholeness that preceded the loss of the preoedipal identification with the mother and the inauguration of subjectivity (Eilberg-Schwartz 1994).

Reformulating Freud's oedipal theory, Lacan—in Eilberg-Schwartz's gloss—casts the phallus as the substitute symbol of desire for the mother and for a wholeness that can never be realized after the child's psychic differentiation from the mother and his/her entry into culture. Why the phallus comes to substitute for the mother's body as the symbol for wholeness is not clear to me. Lacan asserts that we are all, in fact, castrated: "To be human is to be castrated, and men and women share the lack of a phallus" (Eilberg-Schwartz 1994, 28). Is castration the condition men's "nakedness" makes unmistakable, the repressed condition men's love-hate/desire-repulsion for fathers and sons threatens to unleash, that is contained and "reversed" in the curse? How might we reconstruct this traumatic past so that we might dispel the curse? It is a curse that not only relocates the body of the invisible father onto that of the enslaved son, but, as well, onto the embodiment of women (see Bordo 1993). How might (white) men regress through the trauma of castration and (re)member the (maternal) body?

CHAPTER 2

Inside the Tent

"(BE)HINDSIGHT"

A sodomitical impulse was an inherent potential of all fallen male descendants of Eve and Adam. (Jonathan Katz 1994, 49)

[W]hiteness works best . . . when it attaches itself to other abstractions, becoming yet another invisible strand in a larger web of unseen yet powerful cultural forces. (Mason Stokes 2001, 13)

[I]t is by no means accidental that even the linguistic root of our word *masculinity* is the *anus*. (Paul Hoch 1979, 97)

Freud's theories, Lee Edelman points out, postulate a psychic experience in which the key, indeed, constitutive, moments of life history are those that can never be viewed "head on, those that can never be taken in frontally, but only, as it were, approached from behind" (1994, 267). Calculatingly, Edelman quotes Mary Ann Doane's observation that "the psychical layer Freud designated perception-consciousness is frequently deceived, caught from behind by unconscious forces which evade its gaze" (1991, 105; quoted in Edelman, 1994, 267). As Lacan commented (after reading Freud's works of the 1920s—particularly *Inhibitions, Symptoms, Anxiety* [1926]): "it is to the difficulties of recollection we must always return if we want to know where psychoanalysis came from" (quoted in Lukacher 1986, 154). It is to the difficulties of recollection we must return if we are to re-enter the tent.

Presumably, psychoanalysis enables patients to reconstruct earlier experience so that the past's inhabitation of the present can be reconfigured through its remembrance and articulation, in other words, through deferred action. For Freud, human sexuality provided the most defining site in which the effects of deferred action, or *Nachtraglichkeit*, come into play. Alice Pitt (2003, 101) characterizes *Nachtraglichkeit*, as "the time of self-difference and self-resistance." It is "the interminable undulating force of *Nachtraglichkeit*" that intrigues Pitt (2003, 96), its movement in psychoanalytic life-history making, movement enabling us to notice "how we find and lose sight of our capacity to apprehend what matters most to us: the surprise of intersection between our movements onward and our detours back." This is the same temporality—and movement—that structures the curriculum, as Dwayne Huebner (1999 [1967]) appreciated forty years ago.

Ned Lukacher (1986, 35) defines deferred action (*Nachtraglichkeit*) as "a mode of temporal spacing through which the randomness of a later event triggers the memory of an earlier event or image, which might never have come to consciousness had the later event never occurred." One casualty of this notion is any concept of linear causality that works in one temporal direction only. Although the preceding (and presumably imprinting) event is the cause of the later event, the earlier event becomes an effect of the later event. "Rather than offering a simple division between causes and effects," Lukacher (1986, 35) explains, "Freud confronts us with causes that are also effects and effects that are also causes. The random seriality of events that precede and follow the wolf dream leads Freud to posit a double logic of causality that repeatedly turns back upon itself."

Lee Edelman is struck by psychoanalysis' refusal of any conception of a unidirectional temporality for psychic development. Psychoanalysis troubles the logic of the chronological and, epistemologically, any certainty in the relation between cause and effect, as Edelman (1994a, 268) notes that psychoanalysis can be understood as a form "metalepsis," namely the rhetorical substitution of cause for effect or vice versa. Such "substitution," he continues, troubles the relationship between "before" and "behind" (Edelman 1994a, 268). Edelman plays with this spatial—and sexual—image, as he coins the word "(be)hindsight" (1994a, 268) to denote this metaleptic structure in which causes and effects revise each other. The complicity of this structure in the "sodomitical encounter" is discernible, Edelman (1994a, 268) asserts, in Freud's theorization of the primal scene of his patient known as the Wolf-Man.

Bringing into, as Freud puts it, "full view" the "behind" of the present seems especially revealing to Edelman (1994a, 270), given the Wolf-Man's diagnosis, namely an anal-erotic fixation and, concomitantly, an intellectual tendency toward doubt. Freud is trying to account not only for the Wolf-Man's preference for heterosexual relations in which he penetrates his partners from behind, but also for his incapacity for bowel movements, unless produced by enemas administered by male attendants. Freud associates this anal fixation with the skepticism with which the Wolf-Man first resisted his socialization into Christianity, interpreting that skepticism as expressing the ambivalence of the Wolf-Man's erotic attachment to his father (Edelman 1994a). The patient's dream of wolves Freud interprets as the Wolf-Man's "deferred understanding of the primal scene" (quoted in Edelman 1994a, 270), an understanding in which the infant's observation of his parents sexual activity becomes internalized and, then, aggressively directed at himself.

The Wolf-Man's religious skepticism focused on whether or not Christ had a "behind" (quoted in Edelman 1994a, 270). "We catch a glimpse," Freud writes, "of [the Wolf-Man's] repressed homosexual attitude in his doubting whether Christ could have a behind, for these ruminations can have had no other meaning but the question whether he himself could be used by his father like a woman—like his mother in the primal scene" (quoted in

Edelman 1994a, 270). Was there for the Wolf-Man some parallel between his (fantasized) penetration by his father and Christ's abandonment by his Father on the cross? In his mind, were both men "fucked"?[1]

The patient's religious skepticism, Edelman suggests, conveyed his anxiety about his own desire to be stimulated from behind, a desire which subjected him to the law of castration. Edelman emphasizes Freud's phrase—that we only "catch a glimpse"—of this desire and its repression by looking at the primal scene itself through "(be)hindsight." By approaching the primal scene from "behind," Edelman (1994, 271) argues, Freud both resists and reinscribes a "disorienting" inability to distinguish between "outside" and "inside," between what happened and what is remembered, between analyst and analysand.

In what Lukacher (1986, 27) characterizes as the "most dazzling interpretative *tour de force* of his career," Freud derives from Pankejev's dream of the wolves sitting in a tree a primal scene of coitus *a tergo* which he claims the one-and-a-half-year-old boy witnessed one summer afternoon at the hour of five. Although the Wolf-Man never remembers this scene, Freud and his patient are certain that *something* happened prior to the Wolf-Man's dream that accounts for its intensity and its lasting reverberation. The two never do agree on precisely what that event was or how it affected the dream (see Obholzer 1982, 35). Lukacher (1986, 27) observes: "The primal scene explains the wolf dream but has not caused it and is not present in it."

The epistemological status of the primal scene is interwoven, then, with its sexual content. In Freud's reconstruction of the primal scene, Edelman (1994a, 272) underscores, the pregenitally focused infant son perceives that sexual intercourse occurs at the site of the "anus." The primal scene is, then, Edelman (1994a, 272) asserts, "always" apprehended as "sodomitical," specifically as it occurs between sexually indistinct participants who both appear to possess the phallus. Through Edelman's sophisticated theorization we are returned to the sodomitical scene inside Noah's tent.

The primal scene, Edelman suggests, presupposes anal intercourse, but it is not the anal penetration of the father on which Edelman is focused. It is the father's penetration of the son. Is *that* why Noah was enraged? Recalling Lot's seduction (also in a drunken state) by his daughters after the destruction of Sodom and Gomorrah (Gen. 19:31–32), we might ask: was Noah furious because his son seduced him into becoming the penetrator of his own progeny, making a mockery of paternal agency (understood, in this instance, as sexual restraint) and in so becoming, "castrating" his son, relocating him outside the lines of patriarchal geneology and generativity? Was the curse, then, an embittered and repudiated reiteration of the father's sexual subjugation of the son?

In Freud's schema, only later does the son, "painfully" and with "difficulty," as Edelman (1994a, 272) underscores, suppress his identification with the so-called passive position in the primal scene. In psychoanalytic theory, this resolution of the negative oedipal complex is thought to occur in adolescence, accounting for the intense fraternalism of young men (see Young-Bruehl

1996). In world-historical terms, it takes centuries before the slave's enforced (civic) passivity is legally ended. By the disavowal of anal desire, the son protects his narcissistically invested penis from the fate—as he imagines it— of the castrated penis of the mother. In this imaginary turn-around, the penis becomes over-invested; it becomes the phallus. Is this, too, the "curse of Ham," the son fated to *become* the father? Is compulsory heterosexuality also a form of engendered enslavement to which Noah's progeny are now condemned?

Because the participants in the primal scene are not, at least to the preoedipal son who watches them, differentiated sexually, he assumes that both parents possess the penis, as he himself does. In such a scenario, Edelman notes, it is small wonder that the son identifies with each of his parents' positions. Later, after the son makes his identificatory commitment to the father, he experiences (possibly unwelcome) traces of the sodomitical fantasy, even during scenes of heterosexual coupling in adult life. Edelman points out that Freud seems to have "forgotten" the anal character of the primal scene, relegating homosexuality to an unconscious status in the psychosexual development of "heterosexual" men. At least it is there, everywhere, waiting to "come out."

Edelman's interpretation resonates with Aron's (1995; see Benjamin 1995) reconsideration of Melanie Klein's view of the primal scene. Here the primal scene is defined as a field of multiple identifications. Rather than a single same-sex identification—son with father, daughter with mother—the accomplishment of a more differentiated identificatory positionality in relation to the primal scene would not compel heterosexual, genital complementarity. Rather, for Aron, identificatory differentiation suggests the achievement of intersubjective triangularity (see Benjamin 1998).

The primal scene confronts the child with identificatory complexity; she/he must decide with whom to identify, whom to be, and whom to have, not mutually exclusive acts. In the oedipal phase, as Jessica Benjamin points out, this may well require the following series of calculations: If I *am* X, I *love* Y and conversely, if I *love* Y, I *am* X. Or, if I *am* Y, I *love* X, and conversely, if I *love* X, I *am* Y. Binary logic *is* heterosexual logic. But from an intersubjective perspective, recalling Aron's theorization, in the triangular scene the child represents him- or herself as subject *and* object, participant *and* observer, but *outside* the relationship between the two others (the parents). What is crucial for the infant, in Aron's logic, is that each parent is differentiated. "The intersubjective capacity at stake here," Benjamin (1998, 63) explains, "is that of both participating and observing in the same relationship." Was Ham a "participant observer" inside his father's tent?

WINE, FIRE, PHALLUS

Faced with the complexity of racial servitude, reformers found it easier to treat bondage as the sexual condition of white men. (Russ Castronovo 2001, 96)

Oddly enough, while the bible celebrates loving the father, sex with him is anathema. Why? Why does the son's love of the father not issue in incest with the father? (Regina M. Schwartz 1997, 108)

[W]hiteness [is] . . . narcissism—that psychological condition whereby a consideration of "the other" is first and foremost a consideration of "the self." (Mason Stokes 2001, 53)

There is no culture without a drug culture. (Avital Ronell 1992, 96)

Was Noah just "partying" when he got drunk that night? Or was he creating culture? After Lévi-Strauss, Howard Eilberg-Schwartz (1994, 255 n. 17) characterizes fermented beverages in general as a "symbol" of culture. After Bailey (1989, 161–62), he notes that wine has been, on occasion, positively characterized in biblical literature, for instance when it is regarded as a symbol of the blessed age to come (Hos. 9:10; Amos 9:13–15). Even so, excess is condemned. Viniculture, however, is depicted as a civilizational advance. In the passage on which we are focused, Eilberg-Schwartz (1994, 256 n. 17) finds no evidence that Noah is to be "condemned" for his actions.

H. Hirsch Cohen provides details concerning the cultural status of wine in the ancient East Mediterranean world, where it was considered more than a beverage which eased fatigue and warmed the heart. For the ancient Greeks wine, he tells us, fire, and phallus formed the triad which was sexuality. The triad appears on a Grecian vase where on one side Dionysus stands, cantharus in hand, holding a vine of grapes; on the other side sits Hephaestus on an ithyphallic ass. A wine pitcher hangs from the ass's phallus. In another representation of the ass, Hephaestus rides the ass, which has a wine pitcher on its phallus; in another, a naked woman, a maenad perhaps, rides an ass with a wine pitcher attached to its phallus (Cohen 1974).

Such scenes painted on the Grecian vases suggest, Cohen argues, that the wine pitcher is placed where it is because the artist is indicating that wine and phallus are, in symbolical terms, identical. In fact, wine, sex, and fire are the same process, even when the elements are reversed (Cohen 1974). Of course, not all depictions of ancient Greek sexuality were what we would term "heterosexual." Moreover, the association between homosexuality and intoxication persists to the present day. Especially at the end of the nineteenth century, intoxication—specifically that related to drug use—was suggestive of homosexual desire (see Sedgwick 1990, 171–172; Pinar 2002a); it was even compared to semen, the "most important liquor" (see Laqueur 2003, 305).

Associations between fire (or light) and sex were not original to the ancient Israelites. The ancient Egyptians depicted iconographically a connection between sex and fire (light). In the tombs of Ramses VI (middle of the twelfth century B.C.E.), Cohen tells us, a large figure stands—his penis fully erect—with his body and head in the heavens. A series of dotted lines connects his body to heavenly bodies and to twelve little figures holding out their hands to receive little red balls along these dotted lines. One of the dotted lines leads from the tip of his penis to a figure catching a red ball of light.

Cohen (1974) interprets the stream as representing both fire and life, in the sense of birth. It is apparently unthinkable that the figure "catching" the man's *élan vital* is another man.

Cohen's reconstruction of the ancient association of fire (or light) and sexuality takes a masturbatory turn. He imagines "primitive man" for whom "rubbing produced a pleasurable glow . . . in the sexual act" (1974, 6). From penis to dry (morning?) wood, "primitive man" externalized his masturbatory experience, eventually reproducing the "same glow" by "gently pushing a wooden stick through a groove" until it ignited. [T]he "resultant fire," Cohen (1974, 6) imagines, represented a "manifestation" of the "glow" he experienced during "sexual release."

For Cohen—working to establish a context in which we can interpret the story of Noah's drunkenness—these myths reveal "primitive man's" belief that "the glow" during sex, reminiscent of the warmth he felt sitting before a fire, must originate in an identical fiery substance located in the male or female genital tract, that part of the anatomy where sexual "warmth" was felt more intensely. This is, Cohen suggests, the same idea portrayed in the ancient Egyptian tomb as "little red balls as seminal fire" (1974, 7), following along the dotted lines that led from the erect phallus of a God. It is, he continues, the same idea visible on the Grecian vase where the fire of a lantern hung on the phallus of the ithyphallic ass. This notion that fire was located in the genital tract persisted through the eighteenth century, when a French physician designated spermatic fluid as a "fiery substance" (quoted in Cohen 1974, 7).

"Primitive man" imagined that "this seminal fire alone engendered life" (Cohen 1974, 7). Cohen concludes that "prescientific man would have sought to insure himself against its loss [i.e., semen] in the act of procreation" (1974, 7). Surely this was one motive for the Leviticus prohibition of male–male sexual activity. Given that such activity evidently had to be prohibited, we can assume it was not exactly uncommon, and so, returning to Cohen (1974, 7), the "problem" of "replenishing the fire" ejaculated through the penis was "solved" when "prehistorical man" discovered a "fiery substance" that he could drink, and in so doing restore the "seminal fire he lost through intercourse." That drink was alcohol (Cohen 1974). Among the Sambia, the "drink" was semen.

"Prescientific man" (who, it seems to me, is proceeding somewhat scientifically in Cohen's imagination) "observed" that alcohol produced a warm glow in the stomach, a "glow"—the words are Cohen's—which then quickly radiated through the entire body. "Like the power contained in spark or seed," Cohen (1974, 7) suggests, alcohol concentrated "great power" in relatively small amounts. He uses the term "fire-water" to underscore the visuality of alcohol, suggesting that it can "burn the tongue" and "flame up" when "ignited" by a "spark" (1974, 7). Cohen returns to his Grecian vase, where, he notes, the wine pitcher—here a symbol, he suggests, for all alcoholic beverages—is substituted for the lantern on the ass's phallus. The association between the "fiery drink" and the genitalia is, Cohen (1974, 7) concludes, "obvious."

It is this special power ascribed to wine that suggests to Cohen a justification for Noah's intoxication. It was, he offers, no "deficiency of character" that explains Noah's state (a curiously Protestant view of the matter) but, rather, Noah's appreciation that wine could replenish his supply of semen and thereby allow him to execute the command he received from God upon disembarking from the ark. Recall that when Noah left the ark with his family and all living creatures, he built an altar and there burned an offering. God promised never again to destroy all life on earth. He blessed Noah and his sons with what, as Cohen points out, must be regarded as more of a command (a curse?) than a blessing: "Be fertile and increase, and fill the earth" (Gen. 9:1).

God was not beating around the bush: He wanted Noah and his sons to begin the job of replenishing the earth with the human species. (There seems to be no mention that heterosexual incest is involved here, given that the sons had to sleep with their mother to do so, unless they themselves were hermaphroditic. Even this, I suppose, constitutes a case of incest.) After observing God destroy the world, except for them and the occupants of the ark, these men must have taken this responsibility of repopulating a decimated earth very seriously. Curiously, as Cohen notes, God's order was carried out only partially: "These three were the sons of Noah, and/From these the whole world branched out" (Gen. 9:19).

This sentence, Cohen reminds, directly precedes the section describing the drunkenness of Noah. It should anticipate what transpires in the verses to follow. Instead, these sentences introduce the subject of procreation with a summary of its results. Significantly, Noah's name is missing. We are told that the whole world would branch out from the three sons of Noah—is their mother the *only* woman?—not from Noah *and* his three sons. Cohen concludes that Noah failed to carry out God's wish, even though the command to repopulate was issued to all four. Can we say, then, that Noah failed as a "man"?

That Noah failed does not mean, Cohen points out, that he did not *try* to comply with God's command. Cohen (1974, 8) thinks he made a "stupendous effort." Perhaps he failed due to his age; Cohen reminds that he was then six hundred years old. Whether or not Noah's advanced age was an issue, Cohen speculates that he would surely approach his task with resolve, that is, making certain that his procreative capacity was at its maximum strength. To shore up his supply of semen, Cohen speculates, Noah decided he would need wine, and in considerable quantity. To follow the command of God-the-Father, Cohen (1974, 8) continues, Noah planted a vineyard in order to produce that "fiery substance" so as to "increase . . . his seminal fire" and thereby "enhance his generative capacity." But it would appear that the old man overcompensated: "[a]nd he drank of the wine, and became drunk, and lay uncovered in his tent" (Gen. 9:21). Now, there are many interpretative possibilities (including mine, which is that he is disrobed in a gendered sense) here, but let us stay with Cohen.

One rabbinic homily, Cohen reports, divined Noah's intent as conjugal, that is, to have sexual intercourse with his wife, in its understanding of the phrase "in his tent." Because "tent" has the consonant ending generally denoting the feminine gender, the rabbinic interpreter understood it as meaning "her tent," namely the tent of Noah's wife; Noah went to his wife's tent to cohabit with her (Cohen 1974, 8). Indeed, "the 'tent' is the prototypical space of the female," Daniel Boyarin (1997, 144) reports, the "epitome" of "private." Working from these connotations, "tent" could mean Noah's own "feminine self" or, even, his anus. Cohen is not about to go there.

Instead, Cohen links the drunkenness of Noah with the drunkenness of Lot and the destruction of Sodom and Gomorrah. Recall that only four people escaped: Lot, his wife, and his two daughters. Fleeing Sodom just before its destruction, Lot and his family reached the town of Zoar, where Lot's wife was turned into a pillar of salt when she ignored the prohibition against gazing upon the scene of destruction. Fearing that Zoar was not safe either, Lot and his daughters fled to the hill country and finally found refuge in a cave (Gen. 19).

Lot's older daughter felt, evidently, that the whole world had gone up in smoke. The thought that she, her father, and her sister were the sole survivors prompted her to suggest to her younger sister an incestuous plan to save the human species from extinction, obviously a recurring theme in Genesis:

> Our father is old, and there is not a man on earth to consort with us in the way of all the world. Come, let us make our father drink wine, and let us lie with him, that we may maintain life through our father. (Gen. 19:31–32)

Having agreed on the plan, the daughters plied their father with wine until he drank himself into such a semi-conscious state that he was, apparently, unaware that he had sex with his older daughter. The following night the scene was repeated with the younger daughter.

Lot's experience replicated Noah's in a number of significant details, Cohen tells us, missing the rather obvious difference between the two episodes, namely that it was Lot's daughters who "raped" the patriarch and in Noah's case it was, presumably, his son. But Cohen is thinking that Lot, like Noah, survived a disaster of cataclysmic proportions; he, too, believed that he and his children were the sole survivors on earth. Cohen offers that Lot, too, was considered to be an old man at the time of his escape from catastrophe. His age was different from Noah's in years only. Lot also became intoxicated to the point of being vulnerable. The resemblances between the two passages, Cohen (1974, 9) concludes, points to parallel reasons for Noah's "drunkenness" and Lot's "intoxication." One must suppose heterosexism is at work here, as the situation of the two biblical figures seems different in one rather significant detail. It's Lot's daughters who seduce him; it's Noah son who has his way with him.

A Very, Very Hard Thing

For Freud the connection between obsessional neurosis and religious practices was clear. (Sander L. Gilman 1993, 146)

But then people have always known, at least since Moses denounced the Golden Calf, that images were dangerous, that they can captivate the onlooker and steal the soul. (W.T.J. Mitchell 1994, 2)

[C]ircumcision is often a rite that symbolically rips a boy out of the world of the mother and brings him into the world of men. (Howard Eilberg-Schwartz 1994, 160)

It is . . . primarily by looking that we speak our language of desire. (Kaja Silverman 2000, 101)

In Moses' time, boys were circumcised at puberty or in early adulthood; the "surgical tool" was a stone blade. After Joshua led the Israelites across the river Jordan, God commanded him: "Make yourself flint knives and squat down and circumcise the people of Israel for a second time. So Joshua made flint knives and circumcised the people of Israel on the hill of foreskins" (Joshua 5:2–3). Those circumcised in this group ritual were not infants but, as in Egypt, adolescents or young men.

In the Old Testament, circumcision is regularly identified with brutality and occasionally with death. To illustrate this point, Gollaher recalls the episode concerning the rape of Dinah. To avenge the rape of their sister Dinah by a Hivite named Shechem (who afterward proposes to marry her), Jacob's sons tell the young prince, "We cannot give our sister to a man who is uncircumcised; for we look upon that as a disgrace." The scheme of revenge they devise is for the prince, along with all the men in his tribe, to submit themselves to circumcision, lured by the promise that afterward the two families and communities will be able to intermarry. The Hivites agree. "Every one of them was circumcised, every able-bodied male. Then two days later, when they were in great pain, Jacob's two sons Simeon and Levi, full brothers to Dinah, armed themselves with swords, boldly entered the city and killed every male" (Genesis 34:1–25). In this episode, is circumcision revenge? Is it the father's revenge for the rape of the daughter, in metaphoric terms his "rib," the rape of his own feminine "self" in a patriarchal culture?

The following episode suggests so. As a condition for permitting David to marry his daughter, King Saul demands the foreskins of Philistine men as dowry: "All the king wants as the bride-price is the foreskins of a hundred Philistines, by way of vengeance on his enemies." After slaughtering two hundred Philistines, David "brought their foreskins and counted them out to the king in order to be accepted as his son-in-law" (I Samuel 18:24–29). Though the text reads "foreskins," this is peculiar, Gollaher points out, because Old Testament writers refrained from explicitly naming the penis. In all likelihood David did not circumcise the slain Philistines but, rather, in a practice common to many tribes, cut off their genitals as trophies of conquest (Gollaher 2000). Is the foreskin the father's trophy of his conquest of the sublimated son?

At some point circumcision was transformed into a neonatal operation, reflecting both ancient Israelite compliance with the covenant and a political interest in distinguishing Israelite males from their uncircumcised neighbors, an interest that, Gollaher (2000, 13) tells us, "grew acute" during the Babylonian exile (578–522 B.C.E.). Here is an early instance of marking the body so it can be serve as a sign of difference, here ethnic and religious difference, difference that would later become racialized.[2] Gollaher suggests that the priests imagined that the permanence of the mark bestowed in infancy would function to prevent Jews, at least Jewish men, from deserting their community. This would seem to be an explicitly political, rather than religious, motive, and one on which Gollaher makes no comment. About this time, Gollaher notes instead, the procedure itself became more radical, removing a larger portion of the foreskin in order to make it less likely that those who were circumcised as infants could later disguise their Jewish identity. In the film *Europa, Europa*, Sorel attempts this very thing, trying to stretch the remaining foreskin to cover the head of his penis, to hide his Jewish identity from his Nazi schoolmates.

After Alexander the Great conquered the Near East between 334 and 331 B.C.E., Greek culture became fashionable. As a mutilation of the natural male form, circumcision violated Greek aesthetics. Moreover, Greeks held athletic contests in which the young male participants appeared nude. The Greeks' sense of modesty dictated that the foreskin should cover the glans. Visible glans in an uncircumcised man was seen as evidence of sexual arousal, a state considered indecent within the arena. To disguise mishaps (i.e. erections), many athletes wore the *kynodesme*, a strand of colored string that looped around the foreskin, closing it tightly over the glans, a version of what Sorel employed in *Europa, Europa*.

The Greek code of "genital etiquette," as Gollaher (2000, 14) phrases it, positioned circumcised Jews at an "embarrassing disadvantage" in the public baths, wrestling matches, and competitive games where men were naked. To compensate, Jewish athletes built a Gentile-style gymnasium in Jerusalem. They also pulled forward their prepuces—etymologically, the pre-penis, akin to the fore skin—to appear uncircumcised. The latter practice annoyed the priests. Gollaher (2000, 14) quotes Josephus, the eminent Jewish historian, who commented on the trend among Jews in the first century: "They also hid the circumcision of their genitals, that even when they were naked they might appear to be Greeks."

Young Jewish men were not succumbing to conformity, Gollaher notes. There was, as Josephus remarks, a Gentile hostility to circumcision. In part, such aversion stemmed from the association of circumcision with castration. For most non-Jews, the foreskin and penis were not sharply distinguished. Of course, they comprise one organ. Many non-Jews had no idea exactly what was removed when the practice was performed on Jewish babies' genitals. Circumcision was among the mysteries of an alien religion, and, as such, the occasion for rumor and speculation. To counter, rabbis began to defend circumcision on aesthetic, not religious grounds, insisting that the foreskin was an imperfection the removal of which was necessary to reveal the body's ideal form. Odd that an aesthetic preference would become, at times, a "matter of life and death" (Gollaher 2000, 15).

After the appearance of Christianity, many rabbis carried circumcision well beyond its biblical origins. If there is a single dominant theme in the rabbinic texts, Gollaher suggests, it is a preoccupation with blood. As an example, Gollaher notes that the classic ritual, including naming the eight-day-old boy, and the naming prayer in *berit milah*, relies on a passage from the book of Ezekiel: "And it is said, 'I passed by you and saw you wallowing in your blood, and I said to you: 'In your blood, live' '" (Ezekiel 16:6). Rabbi Eliezar commented: "it must be that God said, 'By merit of the blood of covenantal circumcision and the blood of the paschal lamb I will redeem you from Egypt. On account of their merit you will be saved at the end of days'" (quoted in Gollaher 2000, 17). Is blood the parallel to Sambian semen?

Semen and blood flowed from the sons to the fathers, a transfer presumably in the interests of the son. In both rituals, such "menstrual" flows enable to son to rejoin—through identification in the Sambian instance, through the obliteration of earthly identity, death, in the biblical one—the father. Gollaher (2000, 17) refers to Lawrence Hoffman's argument (1996) that the Rabbis merged the two biblical concepts of covenant—sacrifice (from Genesis 15) and circumcision (from Genesis 17).

The patriarchal character of Israelite culture was expressed, Gollaher points out, in the distinction between circumcision blood and the blood flowing from women in menses and childbirth. Intensely patriarchal, Israelite men assigned women to marginal roles in Jewish religious life, a status reflected in their exclusion from circumcision, the "central mark of God's covenant" (Gollaher 2000, 18). Although the blood of circumcision—the emission of the boy's penis—became holier through the centuries, the rabbis, citing explicit taboos dictated in the book of Leviticus (15:19–30), characterized vaginal blood as "uncontrolled, impure, and dangerous" (Gollaher 2000, 18). Gollaher (2000, 18) summarizes: The blood of men was about "salvation," that of women about "pollution." So great was men's misogyny that before modern times rabbinic Judaism never devised a covenant ritual for young women. For most of Jewish history, women's relation to the covenant was derivative, its central symbol reserved for fathers and sons (Gollaher 2000).

The Midrash—Gollaher (2000, 19) characterizes it as an "expansive genre" of rabbinic commentaries on scriptural texts—contains long passages on the significance of circumcision. Gollaher quotes the Midrash-Nedarim: "circumcision is great since, but for that, the Holy one would not have created his world" (3:11; in Gollaher 2000, 19). This "far-fetched" (Gollaher 2000, 19) interpretation is evidently derived from a single passage in the book of Jeremiah, in which God presumably said: "But for my covenant by day and night I would not have set forth the ordinances of Heaven and earth" (Jeremiah 34:27). In other sources, Gollaher reports, numerous tales are told of patriarchs, from Adam to Job, who were born circumcised.

The "most thoughtful and articulate" of Jewish commentators on circumcision, Gollaher (2000, 19) judges, was Moses Maimonides, characterized by Ivan Hannaford (1996, 100), as "the greatest teacher of the Hebrew world" and "pivotal to the shift in Monotheistic thought after 1200." Born in Spain in 1135, trained in medicine and educated in philosophy, Maimonides moved to Cairo where he became personal physician to Saladin. His magnum opus, *Guide to the Perplexed*, remains, Gollaher (2000, 20) tells us, a "classic" instance of "balanc[ing] faith and reason." Because cutting a baby's penis seemed illogical and risky to him, Maimonides labored to formulate a rationale for circumcision (Gollaher 2000).

"No one," Maimonides asserted, "should circumcise himself or his son for any other reason than pure faith." How are faith and mutilation related? Such a question would not have occurred to him; he accepted that circumcision was an indispensable part of Jewish law, and that the procedure had a beneficial effect on men, enabling them to obey the law. Is that because, now mutilated, he is forever reminded that the Father could kill him? The overarching purpose of the law, he wrote, was "to quell all the impulses of matter" (quoted passages in Gollaher 2000, 20). Imagined as "matter," fantasized as enfleshment itself, black Africans would indeed be "quelled."

Maimonides imagined a different consequence. Mutual love and the bonds that bind followed from circumcision, he reasoned. The mark of the Father not only made men's bodies the same, but in so doing obscured self-same sexual difference. It served as a constant reminder of men's spiritual sameness as descendents of Abraham and heirs to the covenant. Only devotion to God-the-Father could persuade a man to undergo such an operation, for, Maimonides allowed, "it is a very, very hard thing" (quoted in Gollaher 2000, 20). Indeed, Gollaher observes, it was fear and pain that prompted the practice of performing the operation on newborns. Maimonides acknowledges that unless circumcision was performed in infancy, many Jews would eschew the practice. Few grown men, he surmised, would willingly undergo such painful procedure (Gollaher 2000).

Maimonides was mindful, Gollaher tells us, that parents would tolerate circumcision only if they managed to deny the pain it caused their sons. At the same time, Maimonides believed that "the bodily pain caused to [the penis] is the real purpose of circumcision" (quoted in Gollaher 2000, 20). On that point, he sounds Sambian. Maimonides believed that the blood, the pain, the very violence of cutting off the skin covering of the penis, represented a trauma that permanently diluted a man's sexual appetite and, moreover, dulled the pleasure he derived from sexual intercourse. "With regard to circumcision," Maimonides wrote, "one of the reasons for it is, in my opinion, the wish to bring about a decrease in sexual intercourse and a weakening of the organ in question, so that this activity be diminished and the organ be in as quiet a state as possible" (quoted in Gollaher 2000, 21). By "suppressing" pleasure and thereby "fleshly temptation," Gollaher (2000, 21) summarizes, circumcision "promoted spirituality." This European cultural logic sets up Africans as "a potential source of fleshly temptation," and, as such, eligible to be "suppressed."

The rabbis knew that the foreskin heightened sexual experience, Gollaher tells us. Maimonides termed it common knowledge that "it is hard for a woman with whom an uncircumcised man has had intercourse to separate from him" (quoted in Gollaher 2000, 21). Was this common knowledge among women, or among men who imagined women unable to resist uncircumcised men? (Or was it that men could not resist uncircumcised men?) Whichever was the source of this "common knowledge," it allowed the sages, Maimonides in particular, to argue that by surgically inhibiting sensuality, making intercourse less pleasurable and more functional (i.e. associated with reproduction), circumcision helped alleviate men's obsession with sex, thereby serving the "spiritual purposes of castration" without destroying men's "fertility" (Gollaher 2000, 21).

We see the same psychosexual dynamic here that was at work inside Noah's tent. The sensual sexual son is suppressed; the compliant sons, castrated through sublimation and condemned to reproduction, are rewarded. When we remember retroactively the genocidal consequences of the curse of Ham, we might also remember that the curse is, literally, upon Europeans and, especially, it seems now, upon European Americans, among the cultural descendents of Ham. It is the European Americans who struggle still within the binaries of flesh and spirit, mind and body, reproduction and degeneracy, compulsory heterosexuality and suppressed homosexual desire. Not until that cultural heritage is dispelled can white racism fade, can a renaissance of European-American culture begin (Pinar 2002b).

Somehow there is, for the Father, a crucial, say we shall, imprinting distinction between voluntary servitude and enslavement, the former associated with control and inhibition, that is, sublimation and reproduction, and the latter with self-abandon and sensuality, even self-shattering dissolution, and degeneracy. Temptation exists to enable self-restraint and voluntary servitude. According to the logic of one of the great anatomists of the Italian Renaissance, Gabriello Fallopio, "God must have imposed circumcision so that Abraham and his progeny would concentrate on serving Him rather than the pleasures of the flesh" (quoted in Gollaher 2000, n. 21). The unsteady separation of the two is evident in white fantasies of Africans and African Americans as cursed to serve, enslaved (whites consoled themselves) by their sensuous natures.

DRINK THIS IN REMEMBRANCE OF ME

[T]his is my body, which is broken for you: this do in remembrance of me . . . . [A]s oft as ye drink *it*, in remembrance of me. I Corinthians 11:24–25

Normative Christian culture . . . disallows reference to the sexual member. Leo Steinberg 1996 [1983], 45[3]

The covenant is a covenant of blood. Lawrence A. Hoffman (1996, 103, emphasis in original)

Historically, Gerald W. Creed (1994) tells us, anthropologists have struggled with the study of sexuality, especially with sexual practices their own culture abhorred. Although anthropologists have, in general, espoused cultural relativism in an effort to reduce one's own encultured skewering of one's perceptions of others' culture, there is, Creed (1994, 66) reports, "something" about sexual issues, especially those monotheistic culture condemn as "deviant," that disables us from setting aside our own cultural assumptions and preferences. To illustrate, Creed cites the work of Williams (1936, 158; quoted in Creed 1994, 66), who condemns "sodomy" among the Keraki as an "unnatural practice" and a "perversion."

Other research attempts to maintain the researcher's relativism by ignoring "unpleasant" sexual practices. To illustrate, Creed (1994) cites Evans-Pritchard's reluctance to report homosexual practices among the Azande. Even though he published several articles and books on the Azande, it was only later that he discussed their homosexual preferences. To provide another example, Creed reports that Kenneth Read (see 1980, 184), reflecting on his own fieldwork among the Gahuku-Gama, confessed that more systematic study might have uncovered homosexuality activity.

To understand the political and social significance of sexuality, Creed (1994, 67) argues, it is necessary to "reject" the "public/private dichotomy," and appreciate that the "private" is "political." Creed credits Cohen (1969) with enabling anthropologists to understand this point when he argued that developing states used sexual regulations to gain political control of their citizenry. Since Cohen's work, Creed continues, many studies (Rowbotham 1973; Zaretzky 1976; Ortner 1978) have elaborated the links between sexuality and politics. Cohen (1969, 664; quoted in Creed 1994, 67) had argued that "there is 'something' about sexuality that renders people vulnerable to control through it."

What might this "something" be? Creed (1994, 67) suggests that it may include the tendency to experience sexuality individualistically. To have sex feels "so private and so personal" that if sexuality is controlled, so is the private individual. To the extent that sexuality itself expresses elements of domination and subordination, he suggests, by controlling with whom one can have sex and how, elements of dominance and subordination can be stereotyped and assigned to specific groups or classes, including, obviously, those identified by gender, race, and sexual preference.

Creed regards ritualized homosexual practices in New Guinea as a mechanism of social control that functions to reproduce a system of inequality based on sex and age. New Guinea ethnographers (Kelly 1976; Herdt 1981) have, he tells us, suggested similar ideas but they have not pursued them. In his intriguing study, Creed will attempt to specify some of the ways in which ritualized homosexuality subordinates and controls women and young men in New Guinea. It is not only compulsory heterosexuality that reproduces patriarchy, it seems, so can compulsory homosexuality. As explored later in this book, this point does not escape Robyn Wiegman.

The practice of homosexuality in New Guinea is widespread, highly structured, and socially regulated, Creed notes. There is a common set of beliefs and actions that typify the practice across Melanesia. As noted in

Gilmore's account, the institutionalization of homosexuality follows from the belief that masculinity can be acquired only through strict adherence to a ritualized regimen. This view of masculinity as acquired is in sharp contrast to the view of femininity, a state that follows naturally, presumably, from the anatomical fact of being a woman. In Melanesia, the mark of masculinity is not the blood flowing from a cut penis, but semen spurting from a fellated one. In Melanesia, instead of shedding blood, boys must ingest semen. The ritualization of homosexuality functions, then, to promote the formation of masculinity by transferring semen from the men to those who are not yet men (Creed 1994).

As is the shedding of blood in circumcision, the transferal of semen is accomplished through a highly structured protocol. Although it must be understood in tandem with other aspects of boys' initiation into manhood, the sex act itself is, Creed tells us, highly structured. As Halperin has pointed out, in ancient Greece there were restrictions concerning who may have sex with whom and in what positions. As in ancient Greece, in Melanesia kinship and age considerations structured the event: certain categories of kin were disqualified, for instance, while sex with other categories of male relatives was mandatory. The younger man in the sexual act must always receive the semen from the older man but, as a boy comes of age, he "graduates" to the man's position—a donator of semen—and, finally, becomes married to a woman and fathers children (Creed 1994).

Although Creed's essay focuses on New Guinea, he also examines Malekula, an island in Eastern Melanesia. There, he explains, anthropologists have documented homosexual practices from a relatively early period. Creed cites Layard's (1942) monograph and the posthumous publication of Deacon's (1934) field notes; together they provide a basic record of homosexual practices in the northern area of Malekula known as the Big Nambas. Each report portrays Malekulan homosexual practices resembling those found in New Guinea.

Deacon (1934, 262, 267) reported that until the time a boy puts on bark belt—the sign of "manhood"—he is in a sexual relationship with an older man. Once he is given his belt this relationship ends, and he himself takes on a boy as his lover. Creed notes that neither Deacon's nor Layard's research identifies the ages of boys at the time of their initial homosexual initiation, nor the age at which they are given the bark belt and graduate to "manhood." Nor do these accounts specify the duration of the role of the "inseminator," although, Creed notes (1994, 69), there are passages that suggest that such roles or positions continue throughout adult life (Deacon 1934, 260–262).

Neither Layard nor Deacon is clear about what the participants think about these homosexual practices. Creed tells us that Deacon (1934, 262) suggests that homosexual practices are believed to cause the boy's "male organ" to develop in size and strength; Deacon assumes that this "male organ" is the penis. But research conducted after Deacon's indicates that many believe in an internal "semen organ" that swells up as the semen acquired in homosexual intercourse accumulates there (Herdt 1981, 217; quoted in Creed 1994, 69). It is possible, Creed notes, that the Malekulan "male organ" is likewise imaginary. Although Layard (1942, 489; quoted in Creed 1994, 69) suggests that ritualized homosexual practices constitute "a transmission of male power by physical means," he does not, Creed observes, specify semen as the vehicle of this transmission. In Christian ritual, it is the body and blood of Christ that are specified as the vehicles of transmission.

There are two features of homosexual practice in Big Nambas that depart from New Guinea practice. One concerns the bond between boy and man. In Big Nambas this relationship is often close and monogamous. Second, chiefs—who appear to achieve this status through birth—are entitled to take on many boy lovers, just as they may have many wives. Although Deacon does not explore any further the intersection of "high status," "wealth," and "sexual access" (Creed 1994, 69), he does provide, Creed tells us, a strong description of a typical homosexual relationship. The father of the candidate seeks another man to act as "guardian" to his son. After the arrangements have been made, this older man has exclusive sexual rights to the young man. He becomes the boy's "husband," and their relationship is "very close" (Creed 1994, 70). The boy never leaves his older lover's side; if one of the pair should die the survivor mourns him (Deacon 1934, 261; see Creed 1994, 70).

Among the Big Nambas, Creed continues, from the time a father selects a "husband" for his son until the time the boy becomes a "man," his "mentor" has absolute sexual rights over him; he would become enraged if he discovered another man having sex with the boy. The older man tends to stay close to his boy. Despite this apparent possessiveness, the older man may sell his rights to his boy-lover for a short periods of time, a practice not uncommon in U.S. prisons (see Pinar 2001, chapters 16 and 17). This arrangement, Creed notes (1994, 85), indicates a system of privatized sexual rights which can be sold, or more precisely, rented by a boy's husband for the husband's economic gain. Moreover, the husband enjoys more than strictly sexual rights; among other "service" obligations, the boy must work in his husband's garden (Creed 1994).

Such relative monogamy is apparently absent in such relationships in the Trans-Fly area of Papua Guinea, Creed reports, citing the research by Williams (1936) and Landtman (1927). He find Landtman's work disappointing, as it merely mentions that "sodomy" is practiced as an initiatory means to help youth grow strong and tall (1927, 237; quoted in Creed 1994, 70). This claim is confounded, Creed complains, by the fact that initiation was no single event among the Kiwai. In contrast, Creed notes, Williams's research on the Keraki tells us much more about their homosexual practices, despite his deprecation of them as an "unnatural practice" and "perversion" (quoted in Creed 1994, 70).

Williams (1936, 158) reports that sodomy was not only universally practiced by Keraki men, it was judged essential to the boy's growth. Boys were sexually initiated at the so-called bull-roarer ceremony at about the age of thirteen. On the night of the ceremony the boy is turned over to a young man initiated previously; he introduces the boy to homosexual intercourse. In all cases reported by Williams (1936, 188; see Creed 1994, 70), the older boy was the mother's brother's son or the father's sister's son of the younger boy. After his initiation, the boy is made available to other male villagers or to male visitors, anyone, it would appear, who wishes to have him.

This sounds like a solution to the problem in Genesis 19. Among the Keraki, however, the sex seemed consensual. The initiates live together in a seclusion hut for several months, during which time they are, presumably, growing at a rapid rate, thanks to constant homosexual intercourse. At the end of his seclusion, the boy becomes a "bachelor," and, as such, he is free to associate more freely with the elders (Creed 1994). He exhibits an increased interest in hunting, presumably a sign of impending manhood. Even so, for another year or so the young man assumes the "passive" position in homosexual encounters (Creed 1994, 70).

One might speculate that such "passivity" would ensure emasculation, but the Keraki have "solved" that "problem." Near the end of the period of homosexual initiation, Creed reports, boys participate in a ceremony of lime eating, a ceremony in which lime is poured down their throats. Creed (1994, 71) tells us that the "severe burns" that follow "neutralize," presumably, the "effects" of such homosexual intercourse; they ensure that the boys do not become "pregnant". After this event, a boy's compulsory "passivity" ends; he is now entitled to adopt the "active" role when the next group of boys is initiated into the bull-roarer (Williams 1936:200–203; Creed 1994, 71). Displaying his heterosexism, Williams (1936, 159; quoted in Creed 1994, 71) wrote:

> It is commonly asserted that the early practice of sodomy does nothing to inhibit a man's natural desires when later on he marries; and it is a fact that while the older men are not debarred from indulging, and actually do so at the bull-roarer ceremony, sodomy is virtually restricted as a habit to the *sertiriva* [bachelors].

Pier Paolo Pasolini participated in a similar pattern of homosexual engagement between older and younger men in 1940s and 1950s Italy, especially among young men of the lower classes (Pasolini 1968, 1982, 1985; Greene 1990). While more formalized, the ritual of the Keraki is not alien to the West.

Like Deacon and Layard, Williams focused upon the role of homosexuality in boys' maturation without linking this developmental function explicitly to the transfer of semen. Creed quotes Williams (1936, 204) who speculates that "the real motive is presumably self-gratification and although the idea of promoting growth is actually present . . . we may be sure that sodomy could get on very well without it." Creed (1994, 71) finds this conclusion somewhat "extreme," but credits it as an acknowledgment of the erotic aspects of homosexual practice, often overlooked when studied as institutionalized ritual. As if in reference to Gilmore's (1990, 147, emphasis added) statement that "[y]oungsters are forced to perform fellatio on grown men, *not for pleasure*, but in order to ingest their semen," Creed (1994, 71) points out, sensibly, that "institutionalized homosexuality is still sex," and, as sex, it may well serve a "pleasurable function." In its religious ritual function, is circumcision parallel to fellatio?

CUT AND SUCK

Circumcision, the very mark of the identity of this people, signals that these sons are not engaged in intergenerational strife with their father. (Regina M. Schwartz 1997, 119)

Identity and desire are so completely imbricated that neither can be explained without recourse to the other. (Kaja Silverman 1992, 6)

[T]he real always makes itself known to us through repetition of symptomatic signs. (Thomas DiPiero 2002, 53)

If the "taboo of looking" functioned to thwart homosexual and, in Noah's case, incestuous desire between father and son, would the ritualization of this repudiation in circumcision function likewise? Circumcision not only inhibited the capacity for sexual pleasure generally, it also protected Jews from destructive sexual urges, Gollaher reports. A thirteenth-century French follower of Maimonides, Isaac ben Yediah, wrote at length about such "advantages" which Jewish men enjoyed over the uncircumcised, for example, Christians. When a woman makes love to an uncircumcised man, Isaac fantasized:

she feels pleasure and reaches an orgasm first. When an uncircumcised man sleeps with her and then resolves to return to his home, she brazenly grasps him, holding onto his genitals and says to him, "come back, make love to me." This is because of the pleasure that she finds in intercourse with him, from the sinews of his testicles—sinew of iron—and from his ejaculation—that of a horse—which he shoots like an arrow into her womb.

The two may have sex two and three times a night, day after day, "yet the appetite is not filled" (quoted passages in Gollaher 2000, 22). Like nineteenth-century Southern white men fantasizing about young black bucks raping fragile white ladies, the site of disavowed identification is the "woman." Because "she" exists in the male mind, this is a site of repressed feminine identification, in a culture cut by binaries, of transposed homosexual desire.

In Isaac ben Yediah's fantasy life, it was Christian men who represented lascivious unbridled desire. Due to circumcision, Jewish men are presumably protected from desire: "He will find himself performing his task quickly, emitting his seed as soon as he inserts the crown . . . . As soon as he begins intercourse with [his wife, presumably], he immediately comes to a climax." For her part, the woman "has no pleasure from him." Not only is his wife's pleasure not a concern (assuming, for the moment, that she takes pleasure in sexual intercourse with her husband), his premature ejaculation serves him spiritually. Now he "will not empty his brain because of his wife [and] his heart will be strong to seek God" (quoted passages in Gollaher 2000, 22). In the eighteenth century, Gollaher tells us, certain Hasidic ascetics took this idea even further, suggesting that circumcision converts the pleasure of sexual intercourse into pain.

Introduced earlier (this chapter, "A Very, Very Hard Thing"), Moses Maimonides is regarded as "pivotal" to a fundamental shift in Western thought after 1200, laying the foundations for "rationality, scientific thinking, international law, and comparative cultural anthropology" (Hannaford 1996, 100). Despite the scope of his intellectual project, he found time to offer technical advice on performing the surgical procedure of circumcision.

> The entire foreskin, which covers the glans, is cut, so that the whole of the glans is exposed. Then the thin layer of skin beneath the foreskin is divided with the nail and turned back, till the flesh of the glans is completely exposed. The wound is then sucked till the blood has been drawn from parts remote from the surface thus obviating danger to the child. After this has been done, a plaster, bandage, or similar dressing is applied. (Quoted in Gollaher 2000, 22)

As we will see, such a procedure poses health risks for boys whose mohels have, evidently, employed their mouths to perform other rituals.

Maimonides discussed unusual cases and rare medical conditions. In the unusual instance of a male infant born without a foreskin, the traditional ceremony, conducted on the eighth day, must still occur. Instead of removing the foreskin, the mohel uses a blade to scratch the child's penis to draw blood. Infants born with ambiguous genitalia, including intersexuals and those born with two penises, were to be circumcised as well. Children born prematurely or with illness brought different advice. Unwell infants were not to be circumcised until they recovered, a judgement deferred until seven days had passed, to ensure that recovery was complete (Gollaher 2000).

The Shulchan Aruch, the standard reference for Jewish ritual observance, states that, despite its symbolic significance, circumcision does not make a boy a Jew. The uncircumcised Jew is, by virtue of birth, still a Jew. Despite this ruling, the *berit milah* (covenant of circumcision) or bris (from the Hebrew word for covenant) has remained, Gollaher (2000, 24) tells us, a "central ritual" within Judaism, a "sacred obligation," an "affirmation" of one's Jewish heritage. He quotes one Mishnah commentator:

> Circumcision draws down a level of Divine light which the Jews cannot draw down through their Divine service. The act of circumcision is necessary, because as long as the foreskin in present, the light will not be drawn down. It is only when the foreskin is removed that the light will reveal itself. (Quoted in Gollaher 2000, 24)

Here circumcision is connected to light and visuality: somehow the foreskin is a filter on the eye, blinding the man to "divine" light. The sublimation circumcision confirms enables one to "look" into the "light," the former a sensual and the latter a spiritual apprehension of a bodiless God-the-Father.

The bris is scheduled during the daylight hours of the eighth day of life, regardless what day that is. For two thousand years, the practice of Jewish circumcision followed the same three-step procedure. First was *chituch*, the cutting of the stretched foreskin. Second was *periah*, the complete exposure of the glans of the penis made possible by cutting and/or tearing away all the inner foreskin tissue back to the frenulum. Finally, the procedure concluded, came *mezizah*, when the mohel "sucked" the blood from the cut penis until the bleeding stopped (Gollaher 2000, 25). Gollaher quotes ethnographer Felix Bryk's description of the mohel at work:

> He takes the member by the thumb and forefinger of his left hand and rubs its several times gently to evoke an erection; he then takes hold of the outer and inner lamellae of the foreskin on both sides . . . and draws them down over the glans, pressing them smooth, by lifting his hand upward at the same time and thus giving the member a vertical position. The *mohel* now takes a pair of small pincers in the thumb and forefinger of his right hand and inserts the foreskin into the crack in such a manner that the glans comes to be behind it and the foreskin that is to be cut away in front of it. Then he takes hold of the knife with the first three fingers of his right hand in such a manner that it rests on the middle finger, with the index finger on the back of the knife and the thumb on the handle. With one vertical motion downwards he cuts off close to the plate the part of the foreskin that is before it, which is being held with the left hand. If this has been done according to prescription . . . the foreskin itself is clipped at the tip, resulting in an opening about the size of a pea. (Quoted in Gollaher 2000, 25)

Is Sambian elders' instruction this methodical?

The son's ordeal is not over. To accomplish *periah* and complete the denudation of the glans, the mohel set aside his instruments and used only his long sharp thumbnail. It was only during the closing decades of the nineteenth century, Gollaher explains, that European and American circumcisers abandoned the use of the fingernail, using instead scissors and other surgical instruments.

> Directly after the cut has been made, the *mohel* puts the tip of his thumb nail . . . into the opening of the inner lamella of the foreskin, grasps the foreskin by its tip with the help of both index fingers, splits it on the black of the glans by means of slitting up to the crown of the latter, and shoves the slit foreskin up over the crown of the glans. (Gollaher 2000, 25)

The incisions made, the mohel pinched the foreskin between his thumb and index finger and tore it from the penis (Gollaher 2000).

Mezizah b'peh followed immediately, the mohel opening his mouth to take the bleeding penis, sucking the blood, then, removing the penis from his mouth, swallowing wine that he then spits onto the boy's penis. Next he places the foreskin in a small basin of sand; then, pouring a fresh goblet of wine, he proclaims a blessing and offers a brief prayer. By this time the bleeding has stopped, the boy needs only a simple linen bandage to cover his wound, to cover his cut and sucked penis. In Central Europe and Italy, the nights before a boy's circumcision were occasions for extended revelry. Men and women ate and drank with abandon (Gollaher 2000). What were they celebrating? If it was about the renewal of the spiritual covenant, would not a more somber ceremony be appropriate?

To review: the male child's penis is first stroked and made erect. Next it is cut, then sucked. Is anyone paying attention here? Evidently not, at least not until the nineteenth century. Although opposition to circumcision within Judaism may have mobilized earlier, the earliest formal objection appears to have occurred in 1843 in Frankfurt. There a group of Jewish laymen founded the Society for the Friends of Reform, a liberal group that published a public manifesto questioning the authority of the Talmud and other religious traditions, among them circumcision (Gollaher 2000). Gollaher (2000, 27) tells us that this issue was "perhaps" the most "controversial." *Berit milah*, the reformers suggested, was not a mitzvah—a rite ordained by God—but a primitive vestige of ancient Israelite culture, outdated and without religious point.

While the public response of the rabbinic community was indignation, privately there were rabbinic leaders who agreed. Rabbi Abraham Geiger admitted that: "I cannot comprehend the necessity of working up a spirit of enthusiasm for the ceremony merely on the ground that it is held in general esteem" (quoted in Gollaher 2000, 28). Geiger agreed that the notion of blood sacrifice, once key to the circumcision ceremony, was no longer relevant, and that the ritual was indeed pointless. He was not alone: liberal German rabbis questioned the valued circumcision for several years, although always in private. Circumcision proved "too divisive, too closely interwoven into the texture of Jewish life and thought, to be debated openly" (Gollaher 2000, 28). Loyalty to the Talmud, marriage between Jews and Gentiles: these were not, compared to circumcision, controversial (Gollaher 2000). Why were boys' penises so significant? Why was access to the boy's penis sacrosanct? I cannot help but think of Southern white men's insistence on the "right" to lynch, another rite of castration (see Pinar 2001, 689ff.).

Not only liberal rabbis and laymen opposed the practice of cutting and sucking boys' penises. European Gentiles tried to restrict it legally. In England, for example, the notorious Jew bill of 1753 restricted circumcision and mohels' participation in the ritual. More sustained opposition, however, accompanied the development of what Nietzsche would name as the new religion, science, in particular, the expanding discipline of medicine. Enlightenment epistemologies were, as we will see, structurally visually; medicine was no exception, as Foucault, among others, have pointed out. Increasingly based on anatomical observation, medical practitioners began to "cast a cold eye" on what seemed, in rational terms, a "risky" and "unnecessary surgery" (Gollaher 2000, 28).

With the emergence of bacteriology, Gollaher reports, the dangers of *periah* paled compared to those of *mezizah*. Even before Pasteur proved the connection between microbes and disease, physicians knew that putting the mouth on an open wound was risky. The appearance of urban hospitals made it possible to document the course of epidemics. Between 1805 and 1806, eight outbreaks of syphilis were linked to infected mohels. In 1833, Krakow alone was reported to have suffered more than one hundred such outbreaks. As a consequence, various states and municipalities attempted to regulate ritual circumcision. In Germany, between 1819 and 1830, for instance, a number of regulations were instituted, among them a requirement that those who perform the procedure undergo specialized training. As well, a physician was required to be present, in case of emergency (Gollaher 2000). The question persisted: were male infants born infected? Or, had the mohels been infected elsewhere?

Evidence began to accumulate that syphilis and tuberculosis—two of the most feared infectious diseases in the nineteenth century, Gollaher reminds— were spread by mohels sucking the blood out of infant boys' cut penises. Jewish physicians conferred with their communities' religious leaders. At first, Gollaher reports, they encountered a sense of helplessness, but in the 1840s this changed, primarily due to an outspoken conservative rabbi in Hungary. Moses Sofer proclaimed that *mezizah* could be abandoned; it had never been, he judged, key. It was, rather, an invention of cabalists who embraced the notion of *mamtik ha-din* ("mouth and the lips sweeten the Law") (quoted in Gollaher 2000, 29), a slogan contemporary gay men may wish to adopt. Of course, many mohels considered *mezizah* not as an act of pleasure, but, rather, as a hygienic measure that stopped the bleeding and cleansed the wound. In many places, mohels who declined to suck the wounded penis were criticized. Despite these attitudes, *mezizah* gradually disappeared as an element of the circumcision ritual, especially in urban centers (Gollaher 2000).

In Eastern Europe, in Russia, within islands of orthodoxy in dozens of countries from Germany to the United States, older men continued to suck the blood from the cut penises of young boys. The practice continued throughout the twentieth century. There were mohels who used glass or plastic tubes to avoid direct oral–genital contact, but many others felt that any compromise of the time-honored practice was heresy. In 1994, the New York City Department of Health was reported a case of an infant Jewish boy with the HIV virus. His mother tested negative. Speculations included that he had somehow been infected while at the hospital or during circumcision. When these speculations surfaced in the press, controversy erupted within the orthodox Jewish community regarding the safety of the operation, to child and circumciser alike, in the age of AIDS (Gollaher 2000).

CHAPTER 3

The Specularity of Alterity

A Way of Life

The intentional stare also threatens to collapse inter-generational separations. (Sheila L. Cavanagh 2004, 326)

Is the *ostentatio genitalium* in Renaissance images of the Christ Child in any sense cognate with the phallic cults of antiquity? (Leo Steinberg 1996 [1983], 46)

Worship is . . . a homoerotic ritual. (Lewis R. Gordon 1996, 248)

To the west of the Keraki, Creed reports, are the Marind-anim, a coastal group studied by Van Baal (1966). The Marind occupy a vast territory along the southeastern coast of Irian Jaya. They illustrate, Creed continues, what Wagner (1972, 19; quoted in Creed 1994, 71) termed the "flamboyant coastal cultures." Between the ages of seven and fourteen, boys engage in anal intercourse (Baal, 1966, 147; Creed 1994, 71). Boys are to be responsive to the sexual desires of their appointed mentors, usually their mother's brother; as well, they are sexually available to any young man who desires them. This arrangement, as Creed notes, combines generalized sexual access with a single older mentor/partner.

The research on homosexual practices on the Plateau, Creed informs us, was done by Kelly (1976, 1977) and Schieffelin (1976), who conducted fieldwork among the Etoro and Kaluli, respectively. Reporting personal communication with the fieldworker Ernst, Kelly (1977, 16; see Creed 1994, 73) explains that Onabasalu initiation is focused on masturbation and the smearing of semen over the initiates' bodies. The formation of Kaluli masculinity is also stimulated by semen. Semen is presumed to have a magical quality that promotes physical growth and mental understanding. In order to become "men," boys must ingest as much semen as possible. Creed (1994, 74) quotes Schieffelin (1976, 124): "When a boy is eleven or twelve years old he is engaged for several months in homosexual intercourse with a healthy older man chosen by his father. . . . Men point to the rapid growth of adolescent youths, the appearance of peachfuzz beards, and so on as the favorable results of this child-rearing practice." Separated from his mother's breast, sucking semen confers manhood.

Creed notes that Schieffelin does not report how much older this "older man" is or what rationale guides his selection, but he (Schieffelin 1976, 126; see Creed 1994, 74) does report that during periods of seclusion in the ceremonial hunting lodge, "homosexual intercourse was practiced between the older bachelors and the younger boys to make them grow, some boys and men developing specific liaisons for the time." Whether or not sex with multiple partners is arranged by the boy's father, Schieffelin is silent. If it is, Creed is curious how the fact of multiple sex partners articulates with the fact of specific sexual relationships. On this issue too, Creed notes, Schieffelin is silent.

To understand the focus on semen in Etoro male initiatory rites, Kelly (1976) treats homosexual and heterosexual practices together, since they both, presumably, represent a transfer of "life force" from one individual to another. In the case of homosexual intercourse, the "passive" boy is presumed to be the beneficiary of the "life force" deposited in him in semen. In the case of heterosexual intercourse, the fetus is presumably the recipient of this life force, as the Etoro believe that the semen deposited in the womb combines with the mother's blood to conceive the child. Kelly (1976, 41; see Creed 1994, 74–75) argued that witchcraft and sexual relations occupy analogous structural positions within the Etoro conceptual system in that both are social practices through which "life force" is transferred from one human being to another (Creed 1994).

Kelly (1976, 52) emphasized, Creed notes, the role of homosexual intercourse in providing boys with ample supplies of semen ("life force") to promote their growth and maturation. To ensure boys become "men," each young man is inseminated through oral intercourse by a single older man from about the age of ten until he has matured and sports a "manly" beard. This occurs in his early to mid-twenties. A ritual of initiation typically takes place during the later portion of this period. Kelly (1976, 47; see Creed, 1994, 75) reports that "youths are initiated into manhood in their late teens or early twenties when they are physically mature (although not fully bearded)."

Approximately every three years during these years of initiation into "manhood," all young male initiates are sent to a seclusion lodge where they cannot be seen by women, thereby disrupting whatever preoedipal ties that remain between mother and son and demarcating physically the presumably separate worlds of men and women. The most recent group of graduates, now "men," stay with the young initiates at the lodge, but, Creed points out, Kelly is silent on the subject of their sexual involvement with the neophytes. Kelly (1976, 47; see Creed 1994, 75) does report that a "generalized insemination of the youths by older men takes place at the seclusion lodge," but the reader is unsure if the previous group of initiates are part of this group of "older men." Although Kelly does not specify the duration of this segregation, he does note that upon reaching "maturity" the Etoro boy-now-man will become the inseminator of a new group of young boys (Creed 1994).

Creed points to Gilbert Herdt's (1981) analysis of the Sambia as the only full-length study of a New Guinea society focused on the institutionalization of homosexuality. Creed (1994, 76) judges this focus "narrow" and as creating the illusion that institutionalized homosexuality is a "system unto itself," focused only on the construction of gender identity, an implication that contradicts, Creed asserts, the fundamental anthropological tenet that culture is integrated. To adequately understand the cultural significance of homosexuality among the Sambia, Creed argues, it must be connected to kinship, politics, economics, etc. The only connection, Creed continues, that Herdt identified is homosexuality and the reproduction of warriorhood. Warriorhood, Creed notes (relying on Meggitt 1977), is successfully reproduced in other New Guinea societies without the aid of homosexual practices.

Herdt conceptualizes Sambian homosexual practices in terms of stages of "masculinization." Like others in the region, the Sambia believe that masculinity develops when young men acquire sufficient semen. Of course, semen is acquired by ingesting it during homosexual intercourse. Around the age of seven to ten, a boy is taken from his mother and subjected to "painful and traumatic rituals" (Creed 1994, 76) designed to drain any residual femininity from the boys' bodies. Now drained of feminine "contaminants," the boys begin to consume copious amounts of semen that, presumably, fill them with "manhood." Fellatio becomes a "way of life" (Creed 1994, 76; see Herdt 1981, 235); the elders require that the boys ingest semen every night. In order to achieve a strong and lasting masculinity, boys must consume enough semen to fill a reservoir (Creed 1994).

When boys' bodies begin to look like men's bodies, older men rule that the boys have ingested enough semen. Now they are promoted to the status of bachelor, a status that entitles them to inseminate a new group of boys. This period ends with heterosexual marriage, often to premenarche girls. The young wife may fellate her husband, but the new couple cannot engage in vaginal intercourse. During this period of about a year or two, then, a young man, Creed notes, is, in practice, bisexual, engaging in oral sex with young boys as well as with his wife. Once the wife's menarche occurs, however, homosexual activities cease and coitus begins (Herdt 1981, 252; see Creed 1994, 77).

As the young man faces fatherhood—the sign of his manhood—his ritual practices shift. Now he moves to defend the manliness he has acquired. Young men begin drinking a white "milk" sap after each occasion of heterosexual intercourse as a means of replacing ejaculated semen (Herdt 1981, 251; Creed 1994, 77). Creed wonders why this sap is not used in the first place as a source of semen for initiates. Creed faults Herdt for failing to look for other possible functions of the ritualized homosexuality, functions in addition to the acquisition of semen. Herdt notes the subordinate–dominant structure of these generationally defined homosexual relationships, but, Creed complains, he does not analyze this structure.

Perhaps ritualized homosexuality among the Sambia functioned to interpellate boys as "men." In accord with object relations theory, Herdt (1981; see Creed 1994, 77) characterizes Sambian homosexual practices as doing just that, inculcating masculinity and furthering the process of male "separation-individuation." "Ritualized homosexuality reinforces the rigidity of the masculine ethic, it allows for no exceptions in the race for acquiring maleness" (Herdt 1981, 322; quoted in Creed 1994, 77). Herdt (1981, 305) employs terms such as "radical resocialization" and "ritualized gender surgery" to describe the "replacement" of the gender identity a boy learned from his mother—a gender identity that prepared him homosexually?—with the normalized masculine gender identity. Creed notes that Herdt does not explain why homosexuality serves as the means of radical resocialization, a strategy other cultures, while still accomplishing the same end, do not employ.

It is evidently beyond the purview of professional obligation to point out that we in the West manage without ritualized homosexuality. We manage, if clumsily, with boys too often feeling abandoned and angry and forced to look to male peers—themselves abandoned and angry and looking for solace and intimacy—to create homosocial networks that simulate homosexual bonding in sublimated but often violent ways. Is the fascistic fraternalism of contemporary homosociality one consequence of homosexual repression and "compulsory heterosexuality"? For Lacan, Kaja Silverman (2000, 122) tells us, "repression [itself] is . . . virtually synonymous with the normative or heterosexual Oedipus complex."

Creed points out that Herdt's "resocialization" thesis follows Burton and Whiting's (1961) explanation of male initiation as "psychological brainwashing." Burton and Whiting (1961) had argued that puberty rites destroy the feminine identity learned by boys during the intense period of mothering, "replacing" it with heteronormative masculine identity. Although such rites, in general, and ritualized homosexuality, in particular, Creed (1994, 77) notes, are "obviously" meant to construct masculine gender identity, homosexuality cannot be limited to this one function. Moreover, Creed argues that Herdt's (1981) "radical resocialization" thesis does not require homosexuality at all. He points out that although the first several years of a boy's life are "dominated" (Creed's word, see 1994, 77) by his mother, during this time he cannot help but observe and experience the gender polarization around him. Despite the symbiotic intimacy with his mother, the boy cannot help but notice the sexual divisions that structure his culture. Creed speculates, reasonably, that boys are probably taught masculinity, not only by their mothers, but by others associated with them and their parents. In monotheistic culture, boys are taught as well by television, film, sport, and school, instilling gendered and racialized economies of visibility.

Are older men "wolves" whose prey are young boys? Is this the other side of the same coin of Freud's primal scene in which the sons murder the father, then consume and worship him as disembodied authority? Or is that a deferred and displaced conception of a prior primal scene, a scene which men in New Guinea enact in timeless repetition? Without proscriptions against viewing the naked body of the fathers, are Sambian practices what would happen had Ham encountered Noah "outside" ancient Israelite culture? Why are these young men not "feminized" and "castrated" by their submissive and "passive" roles in fellatio, as they are, presumably, in the West?

I have focused here on the medium of transmission—semen—between father and son, older men and younger men. I have ignored the site of the transmission, the symbolism of the apertures. In the preoedipal primal scene of Sergei Pankejev (the "Wolf-Man"), there were no totems without apertures: vagina, mouth, anus. In the Wolf-Man's "primal scene," he was opened to his father's desire; he wondered whether Jesus, too, was open to his Father in like fashion. Schreber was opened. Is the aperture that enables men to be open to other men what Noah experienced as "lack"? Is patriarchy the defensive erection of totems to penetrate others' apertures?

A Dreadful Secret and an Indispensable Pivot

That Fathers have a certain amount of this accepted power in the Family means that their sexual abuse of children is not so much a deviation from normal familial relations as an illustration of them. (Vikki Bell 1993, 62)

But what makes the images I am citing rare and psychologically troubling is the Father's intrusive gesture, his unprecedented acknowledgement of the Son's loins. (Leo Steinberg 1996 [1983], 105)

Two forms of taboo desire, incest and interracial sexuality . . . are linked symbolically to the potentially tragic narrative homosexuality. (Siobhan B. Somerville 2000, 122)

"Incest was a popular practice," Foucault (1990, 302) reminds us, "widely practiced among the populace, for a very long time. It was towards the end of the nineteenth century that various social pressures were directed against it. And it is clear the great interdiction of incest is an invention of the intellectuals." As Foucault and others (for instance, see Twitchell 1987) have pointed out, incest is a practice that continues in secret, a secrecy made possible by the organization of the nuclear family, the affairs of which are conducted in private under the jurisdiction of men (Pronger, 1990; Gartner 1999). In volume one of the *History of Sexuality*, Foucault asserts that incest is an innate aspect of family life: it is "an obsession and an attraction, a dreadful secret and an indispensable pivot" (1976, 109; quoted in Pronger, 1990, 65n.).

The taboo against incest is usually presented as if it were instinctual; transgressions are punished as if they were unnatural acts. Social psychologist Roger Brown (see 1965, 751) points out that if the taboo were instinctual there would be no need for the taboo. Perhaps acknowledging this paradox, James Twitchell (1987, xi) characterizes incestuous acts as "*uncultural* acts." Surveying incest historically, Twitchell cites its institutionalization in royal families: Cleopatra, for instance, was the offspring of at least eleven generations of incest and was herself a sibling partner. Regina Schwartz (1997, 99) also points out that "incestuous marriages were institutionalized in the ancient world. And even though biblical law strictly forbids it, biblical narratives equivocate." Writing before the 1990s, fascination with "suppressed memory," which resulted in prominent criminal prosecutions of child-care workers and of parents, Twitchell (1987, 13) nonetheless concludes that "the incidence of incest is simply too common to be ignored any longer. It is approaching the status of epidemic." He suggests that millions of women have experienced what was thought to be rare.

Incest is, it would seem, omnipresent. Twitchell reports that Father–daughter or surrogate father-figure incest is the most prevalent practice (approximately 70 percent), followed by brother–sister incest, including adopted or "rem" siblings (20 percent), and that the remainder are uncle–niece or in-law relationships, and, in much smaller numbers, mother–son. What of homosexual incest? Twitchell (1987, 13) writes: "Incidents of father–son and, especially, mother–daughter incest were supposedly unheard of, but these homosexual liaisons do occur and if their symbolic forms, which we see in sports and advertisements, are any indication, the percentages may be higher than anyone expects." It is the dreadful secret recorded in Genesis 9:23.

Of all these various combinations, Twitchell suggests that the mother–son relationship is the most detested; it, he observes, is that act of incest for which "we reserve our linguistic wrath" (1987, 54). No other liaison provokes language as "obscene and ferocious" as the special curse—not only in English language but almost all other languages as well—that is "mother-fucker." Only the mother–son relationship provokes such "social and familial outrage at a fever pitch" (Twitchell 1987, 54). Why would that be? As object relations theory suggests, the mother–son symbiotic relationship is replaced by the covenant between father and son, marked through circumcision, enabling the son to metamorphose into a "man."

Twitchell (1987, 71) asserts that the figure of the vampire is "the father's ultimate statement of power." He continues: "For a myth so loaded with sexual excitement there is no mention of sexuality. It is sex without genitalia, sex without confusion, sex without responsibility, sex without guilt, sex without love—better yet, sex without mention" (Twitchell 1987, 71). The penetration of the vampire is most popularly heterosexual—the young vir-ginal white girl falls victim—but as *Interview with the Vampire* (1994) makes clear, there are homosexual elements as well.

Biblically, incest appears to be tolerated when reproduction is at stake. The incestuous necessity of Adam and (St)Eve has been understood and forgiven, as has been the situation of Lot and his daughters (Gen. 19:30–38). However, as Twitchell (1987) points out, this imperative to reproduce would not seem to be at work in the half-siblings Abraham and Sarah (Gen. 20:12), or uncle–niece pairs like Nahor and Milcha (Gen. 11:27, 11:29), aunt–nephew pairs like Amram and Jochebed (Exod. 6:20), or the brother–sister rape of Amnon and Tamar (II Sam. 13:2, 14, 28–29). He does not mention Noah sharing his wife with his three sons.

Over the centuries, conceptions of childhood shift (Aries 1962). By the end of the eighteenth century, the child was seen no longer as a deformed adult, but, rather, as an innocent under surveillance, in need of guidance (Baker 2001). This tendency toward adults' fantasizing of the child as innocent continued in the nineteenth century as the romantics repeatedly referred to youngsters as lambs, divine philosophers, the father of man. That century concluded with Victorian novelists portraying children as angels exploited by older villains but prevailing in the end. This historical shift from deformed adult to innocent child was reflected in popular culture (Twitchell 1987). And not only in the late nineteenth century: writing about the late twentieth century, Leo Bersani (1994, 261) observed:

> Adult sexuality is split in two: at once redeemed by its retroactive metamorphosis into the purity of an asexual childhood, and yet preserved in its most sinister forms by being projected onto the image of the criminal seducer of children. . . . More exactly, the brutality is identical to the idealization.

The very conception of innocence is self-corrupting, as Pasolini (see Stack 1969, 124) knew and Lee Edelman (2005) makes clear.

In the late nineteenth century, the conception was codified into legislation. Demanding action, the British National Society for the Prevention of Cruelty to Children flooded Parliament with "moral statistics" as well as with impassioned testimony. In 1908, the Incest Act defined both the act (emphasizing the usually female victim but exempting stepchildren) and the punishment (imprisonment of no more than seven years and no less than three). The variegated social crusade against sweatshops, prostitution, drunkenness, gambling, and other "sins of the flesh" focused upon child molestation. It forced an unwilling and, as it would turn out, largely powerless state bureaucracy to intervene in this unmentionable problem (so it was imagined) of the working classes (Twitchell 1987).

As in the instance of the sentimentalized, mythologized nineteenth-century white woman and the feared black man, this "child" is not real either, a figment in the Western imagination (Baker 2001; Pinar 2006). Twitchell points out, it took men who were not actively engaged in child care to be able to imagine the child as angelic or, at least, innocent. Like their English counterparts, the most aggressive spokesmen of American romanticism—Thoreau, Whitman, and Emerson—were men who had little or no family of their own. These American intellectuals were, Twitchell (1987) complains, a recluse, a homosexual, and an extremely passive parent and distant husband.

Roger Brown likens the prohibition against miscegenation to the incest taboo (see 1965, 751). In 1880, for instance, the Mississippi legislature banned intermarriage, declaring it to be "incestuous and void" (quoted in Sollors 1997, 400). The statute provided the same punishment for miscegenation as it did for incest, codifying a conflation that would be evident in Adolph Hitler who spoke of "incest" when an Aryan interbred with a non-Aryan (Sollors 1997). Freud thought primitive—dark-skinned—peoples were more "liable" to the "seductions" of incest (see Eng 2001, 8). Did these associations of incest with race begin that night in Noah's tent, when a son sodomized a drunken father (or what it the father who sodomized the son? or was it consensual?), who retaliated with a curse that turned his grandson into a servant, in the white male mind, a black slave?

Is father–son incest the primary sin of knowledge that Eve transgenders and Europeans racialize? Is Adam God's rib? From Genesis it would seem that "woman" is a fragment of "man," but, in fact, is not the reverse the case? Is not "man" a chromosomal aberration of woman? Is not the boy his mother during the fetal stage, sharing both body and identity? Does, perhaps, the rape of Noah carry within it a fantasy of all men to return through the womb back to the testicle of their fathers? Is "salvation" the sacred fantasy of such merging? For the male fragment, is not homosexual union self-affirmation? And in evolutionary terms is not the original father from Africa, is he not black? "Was Yahweh . . . black?" (Fichte 1996, 359).

Several nineteenth-century white writers, Mason Stokes (2001, 87) reports, decoded the "serpent" in the Garden of Eden as either "an ape or a human, often black, and usually male. . . . Thus miscegenation becomes the reason for the Fall." In addition to the snake as projective screen, two microcosmic insects also "haunt" the imaginations and populate the projections of men, Elizabeth Grosz (1995, 188) reminds. These are the black widow spider and the praying mantis, two species that have come to represent an "intimate" and "persistent" association between "sex" and "death," between "pleasure" and "punishment," between "desire" and "revenge" (Grosz 1995, 188).

Given late-nineteenth-century obsessions with the disappearance of the "white race" through miscegenation and twentieth-century obsessions with the spell-binding seductive powers of the black phallus, not only female sexuality occupies the projective contents of these species. Despite the color, the mantis, more than the black widow, may evoke men's fantasies: she/he is the predatory and devouring seductress (regardless of race) who ingests and incorporates her mate, castrating or killing him in the process. The mantis, Grosz (1995, 191) observes, is the *"femme fatale* writ small." Like the serpent, the self-shattering ecstasy that possession threatens tempts the white man who must cover up the naked truth.

THE TABOO OF LOOKING

Discipline is a mode of power that works through observation. (Vikki Bell 1993, 63)

[V]ision provides the agency . . . [and] the mechanism through which the male subject assures himself that it is not he but another who is castrated. (Kaja Silverman 1988, 17)

None of you shall approach to any that is near of kin to him, to uncover *their* nakedness: I *am* the Lord. (Leviticus 18: 6)

Recall that, after hearing Ham report that he had seen his father naked, Shem and Japheth took a garment, walked backward—in order to avoid seeing their father naked—into the tent and covered him. The narrator tells us: "Their faces were turned away, and they did not see their father's nakedness" (Gen. 9:23). Cohen concludes that Ham's sin consisted "solely" of gazing upon the naked body of his father.

Shem and Japheth's behavior points to such conclusion, Cohen argues. Had Ham castrated his father, as suggested by the rabbinic sages, the brothers would have attended to the wound, not simply draped a garment over him. By telling us that the sons did nothing more than cover their father, the narrator seems to be saying that this measure was sufficient. That Shem and Japheth approached their father with their faces averted, Cohen emphasizes, suggests that they were determined to avoid their brother's sin. Focusing on the facts of the text, then, Cohen (1974, 15) feels confident that Ham's "offense" consisted of "gazing" upon his father's "nakedness."

"Looking was not," Cohen (1974, 15) explains, "the simple act for biblical man that it is today." (It is not so simple today: see, for instance, Rosen 2000). In ancient Israelite culture, the "look" implied danger. Cohen (1974, 15) cites the following examples: God refuses Moses' request to allow him to behold his presence with the warning that "man may not see me and live" (Ex. 33:20); Manoah, the father of Samson, after viewing the angel of the Lord, cries to his wife: "We shall surely die, because we have seen God" (Judg. 13:22): Elijah avoids the sight of God by covering his face with his mantle; and, perhaps most well known, Lot's wife turns into a pillar of salt upon viewing the destruction of Sodom and Gomorrah.

To explain why the act of looking would have been deemed perilous, Cohen turns to a psychoanalysis (if only for a moment) to suggest that looking implies identification. He quotes Fenichel: "If a man looks upon God face to face, something of the glory of God passes into him. It is this impious act, the likening of oneself to God, which is forbidden when men are forbidden to look at God" (1953, 391; quoted in Cohen 1974, 15). Cohen suggests that Lot and his company were forbidden to look upon the destruction of Sodom and Gomorrah for a similar reason. Even if, he (1974, 15) speculates, the ancients were unfamiliar with the concept of "identification," they may well have concluded that there was "something" in the "act of looking" that "closely resembled" that psychoanalytic concept: by "looking" at someone, one could "acquire his characteristics." Put simply, to look was to acquire (see Cohen 1974, 15).

The Hebrew language, Cohen tells us, reflects the notion that looking can be a means of acquisition. He points to the number of Hebrew words for "see" or "look" that are identical with or closely related to the words for "fence" or "wall." The fence, Cohen asserts, symbolizes ownership: "everyone recognizes that the enclosed object has been acquired at some time in the past and is owned by the person who has erected the fence" (1974, 15). The ancient Israelites, he continues, related the word for "looking" to the word for "fence," thereby indicating that something could be encompassed even without the visible signs of enclosure. In other words, the one who is looking—the eye of the beholder—could encompass everything within his range of vision. In this sense, Cohen is arguing, the Hebrew language conveys the idea that looking can mean acquiring that which is viewed. The idea occurs in English in the words "hold" and "behold" (Cohen 1974, 15).

A fence or wall (in Robert Frost's sense) can also imply separation and distance, allowing for and, indeed, protecting that privacy and inner space that enables one to breathe freely. Frost suggests, of course, that such separation is necessary to good neighbors, not good patriarchs. In ancient Israelite culture, however, to "enter" someone's property, say a tent or a body, was not only to violate another's "property," it was to transgress the "erections" that punctuated patriarchy. If patriarchy substitutes private property for symbiosis (or relatedness), or "having" for "being," then Ham has struck a blow (as it were) for the matriarchal state. His progeny must be enslaved.

Now that Cohen has defined the act of looking as "acquiring," he thinks he can now explain that why Ham left the tent to tell his brothers what he had seen. Ham was, Cohen is certain, not trying to undermine his father's honor by making him the "butt" of his "dirty mouthings" (Cohen 1974, 16), as some commentators contend. Rather, Cohen (1974, 16) suggests, Ham was merely staking his claim: by "looking" at the naked body of his father he "acquired" his father's "potency." This is similar to the analytic tactic historian Joel Williamson (1984) employs in his brilliant if heterosexist attempt to explain lynching, that is by castrating young black men, white men were trying to acquire black men's sexual potency (see Pinar 2001, chapter 19, section II). It is an interpretative tactic that assumes that identification succeeds in its disavowal of desire (see Butler 1997).

Perhaps Cohen senses the instability of "identification," because he moves quickly from the scene of the son gazing upon the naked father. He imagines Noah not as "passive" but as "active," as in heterosexual intercourse. To "claim" his father's "potency" suggests that the son caught Noah in the procreative act, Cohen (1974, 16) reasons. The text, Cohen acknowledges, says nothing about Noah engaging in (hetereo)sexual intercourse, only that Noah was drunk and naked inside his tent. Yet, Cohen insists, somewhat stubbornly, "from the information given, Noah evidently did have intercourse or intended having it" (Cohen 1974, 16). From the information Genesis provides, the only sexual intercourse Noah may have had was with his son.

Cohen (1974, 17) continues his effort to exonerate the patriarch, telling us that Noah's nudity "most likely" was not related with his being drunk. He was naked, Cohen reasons, neither because he was incapable of controlling his actions nor because he was overheated from drinking so much wine. Rather, Cohen suggests, Noah's nakedness was preliminary to sexual intercourse. Recall the relation between wine and potency elaborated in chapter 2, "Wine, Fire, Phallus": Noah drank the wine to acquire the "seminal potency" he would need to repopulate the earth (Cohen 1974, 17). Once sufficiently "fortified" (1974, 17), Noah stripped for sex. The narrator, Cohen (1974, 17) is forced to acknowledge, "seems to be averse to furnishing further details, since he does not proceed further along this subject." So although the text says nothing to even hint that Ham observed his father having sex (returning us to the Wolf Man's primal scene), Cohen (1974, 18) speculates that the son "must have been present" throughout the sex act, looking so as to acquire his father's "strength" through his "gloating stare." "Possessing" his father's strength, Cohen (1974, 18) continues, Ham would thereby emerge as the "most powerful" of the three sons, destined to "inherit" the "mantle" of "leadership" upon his father's death. It is curious, given this reconstruction of the scene, why the other sons were not the ones to become angry. After all, would *they* not have felt competitive, wished *they* had been the ones to "acquire" Noah's potency? Why, in this scenario, would Noah become *enraged*? Should he not have been proud that his son wanted to be "potent" like his dad?

Still immersed in this primal scene, Cohen (1974, 18) tells us that, after all, it is "reasonable" to "infer" that Ham could not simply tell his brothers what had transpired inside the tent. He had to provide documentation. How? There is only one piece of evidence, Cohen decides, that no one could deny: the clothing that Noah had shed preliminary to intercourse. Cohen notes that Ham produced Noah's garment, narrated in Genesis 9:23: "Then Shem and Japheth took a garment, laid it upon their shoulders, and walked backward and covered their father's nakedness."

Cohen notes that the key word in the passage seems mistaken, but that is only apparently so. Instead of "*a* garment" as "any old garment, we must read it as denoting "*the* garment." Though on occasion in the Bible the definite article expresses a general definition, Cohen explains, the definite article in this instance designates something specific, in this case a particular item of clothing. That "specific garment" could have "belonged" to "no other" but Noah, Cohen (1974, 19) continues. Reconstructing the scene, Cohen is sure that "Ham must have skirted the sleeping, naked Noah, picked up his father's garment that had been cast aside, and stepped outside to show '*the* garment' to his brothers" (Cohen 1974, 19). Perhaps the garment Noah shed was his "gender" as "man"?

Shem and Japheth appear to be acting more out of fear than respect, Cohen notes. He speculates that they probably fear being "infected" by their father's debility (Cohen 1974, 20). Or, in a queer reading, perhaps they fear being infected by homosexual incestuous desire? Despite Cohen's heterosexist interpretation, a queer note is audible. Cohen notes that instead of showing respect for Noah, the brothers appear to regard their father as "irreparably weakened" (1974, 21). Is that because he is "spent" after phallic heterosexual intercourse? Or is it because, due to rape, castration, or being "acquired" by his son in the gaze, he is now "unmanned," now a "woman"? Was Noah a curse away from becoming Schreber?

None of these interpretative possibilities occurs to Cohen, who continues to focus on Shem and Japheth walking backwards, avoiding looking at their naked father, to "protect" themselves against possible "infection" (1974, 21). (This passage anticipates Exodus 40:34–35, wherein Moses covers his face in fear to look at God, fear that does not prevent him, however, from peeking through a narrow chink to see the Father's backside [see Schwartz 1997, 116].) Cohen suggests that the biblical narrator evidently regarded the threat of possible infection as key to the event, for she/he stressed the fact that Shem and Japheth could not possibly have seen their father's "weakness": "[T]heir faces *were* backward, and they saw not their father's nakedness" (Gen. 9:23). Perhaps, Cohen (1974, 21) speculates, Noah's "weakness" was the reason he failed to beget any more children, despite the "seminal fire" derived from the wine. Cohen suggests that Noah's generative power, once appropriated by Ham's voyeuristic gaze, was, then, too diminished to function.

And now the founding moment—or so it would be designated later, in "(be)hindsight"—of racialized enslavement in the West: "And Noah awoke from his wine, and knew what his youngest son had done unto him" (Gen. 9:24). There is no suggestion in the text that Shem and Japheth felt they must inform their father what happened to him, Cohen notes. There was no need to tell him. He already knew.

Cohen feels he has solved the mystery of Noah's nakedness, except for one item. "There remains," Cohen (1974, 29) acknowledges, "to be explained the reason why Noah vented his wrath upon the innocent Canaan when Ham was the affronting party." He quotes the relevant lines:

> Cursed be Canaan:
> The lowest of slaves
> Shall he be to his brothers. (Gen. 9:25)

Assuming the text is not "disturbed," that is, does not reflect two different interpretative traditions merged into one, Cohen is ready to resolve this mystery, too. He offers that, in his view, Noah's curse—relegating Canaan to the lowest of slaves—was not as "unjust" (Cohen 1974, 29) as it might seem, especially when we consider what Noah could not do. Cohen assumes that Ham knew that once he possessed his father's generative power Noah would not avenge himself against the now potent son, now able to pass along this potency to his son Canaan and his progeny. "To thwart Ham's scheme," Cohen (1974, 29) speculates, "Noah—if this hypothesis is correct—would have had to curse Ham's son, Canaan, who was not shielded by any such generative power." Even without being "disturbed" by heterosexism, this interpretation seems a stretch.

Cohen is undeterred. "Far from acting out of vengeance," he writes, "Noah seemingly degraded the future generations of Canaan to frustrate Ham's design of transferring his newly acquired special strength and power to Canaan and his progeny" (Cohen 1974, 30). Would not a grandfather want that? That question is never raised, as Cohen continues intently within the logic of his speculation, noting that Noah's decree had to be pronounced before Ham could transfer the force of Noah's potency to his son, thereby making him invulnerable. With Ham's scheme blocked, he was, in this scenario, no longer able to threaten the position of Shem and Japheth and their progeny, a problem never raised before. Cohen (1974, 30) concludes, simply: "Noah assured the safe succession of leadership." Surprise: the reproduction of patriarchy is accomplished through servitude.

A RACIALIZED CULTURE OF VISION

Why visibility as a privileged telos? (Mason Stokes 2001, 160)

No metaphor was in as much need of such an unsettling as the founding trope of Western metaphysics, the privileging of whiteness over blackness, light over darkness. (Martin Jay 1993, 509)

"Light" would then seem to be another name for the value inhering in whatever is first loved, and displacement something like the transport of a candle lit in this primal flame to ever more remote territories. (Kaja Silverman 2000, 16)

Long before Plato—David Levin (1993a, 1) points to fragments attributed to Parmenides (475 B.C.E) but not to Noah—philosophical thinking in the Western world was drawn both to the "authority" of "sight" and fearful of the "dangers" of trusting in vision and the "objects" it reveals. These dangers included not only the "tricks" and "deceptions" of everyday perception (Levin thinks of the stick that appears to be bent when placed in water), but also the "illusions" and "superstitions" associated with "visionary" religion. Discernible in the cultural lifeworld of the ancients, Levin (1993a, 2) reports, were "occult" visionary religions, visionary rituals, and visionary "technologies of the self," this last phrase a reference to Foucault.[1]

Although the earliest church fathers—Origen, Tertullian, and Clement of Alexandria—worried that religious images communicated an "overly anthropomorphic notion of the holy," Martin Jay (1993, 36) points out that their successors were not reluctant to employ the visual to make Christianity more accessible to new believers from non-Jewish backgrounds. As early as the Hellenization of Christian doctrine undertaken by the converted Jew, Philo of Alexandria, in the first century, Jay notes, biblical allusions to the auditory were systematically replaced by those referencing sight. The Gospel of St. John alleged that "God is Light," and medieval thinkers like the Pseudo Dionysus interpreted the phrase literally. Churches built by the converted Roman Emperor Constantine were "filled" with light, Jay (1993, 37) notes, a "residue" of the "earlier cult of the sun."

Levin (see 1993a, 3) reports that there is evidence for a shift from the privileging of seeing to that of listening. Levin cites Hans-Georg Gadamer's appropriation of the ocular concept of "horizon" in his conceptualization of a conversation-based hermeneutics of interpretation. He points to the work of Jürgen Habermas which, like that of John Dewey, replaced the detached-spectator epistemology of scientism with one that emphasizes communication and democratic participation, releasing the modern subject from what Levin (1993a, 3) characterizes as a "terrible double bind." That bind—in it we see Edelman's analysis of "before" and "behind"—follows from the positioning of the subject, in the objectivism associated with ocularcentrism, "either" in the role of a dominant observer "or" in the role of a visible object, "submissive" before the panopticon of power (Levin 1993a, 3). This "double bind" gets performed in certain sexual practices, among them sadomasochistic sexual practices among men (see Bersani 1995). It structures the racial economy of visibility.

By multiplying perspectives, Levin (see 1993a, 4) suggests, Nietzsche turned sight against itself, subverting the authority of ocular thinking, challenging the "visionary" aspirations of philosophy, awakening some of us from that "long dream of metaphysics" (Silverman 2000, 2). Altering the visionary ambitions of philosophy does not, however, necessarily alter the visionary ambitions of mass culture, as society and culture are still being structured by the "hegemony" of "vision" (Levin 1993a, 5). During the twentieth century, Levin points out, three major philosophers—Martin Heidegger, Michel Foucault, and Jacques Derrida—argued that not only has the historical privileging of sight continued, but the worst tendencies of such ocularcentrism have dominated in distinctively modern ways (see also, Jay 1993).

For Heidegger, visionary experience has always dominated both the origin (*arché*) and the end (*telos*) of philosophical metaphysics. But the hegemony of visuality in modernity—what in his "Letter on Humanism" he calls the "malice of rage" (quoted in Levin 1993a, 5)—has become increasingly "enframed" as the dominant *episteme*. Like Henri Bergson, Jay (1993, 270) reminds, Heidegger mourned the devaluation of "temporality in Western metaphysics since Heraclitus" and its replacement by a "spatializing ontology based on the synchronicity of the fixating gaze." This conjunction of visuality and spatiality has meant a tendency toward a certain essentialism in representations of self and others, of knowledge, truth, and reality, an essentialism that distorts the visible world by imposing irreconcilable differences between subject and object. Ham's son is no longer a beloved grandson but an object, a servant whose "descendants" would become slaves. The consequence of "looking" is banishment from the "tent," from subjectivity, condemned to worlds of servitude, in our time, to the surface of the body: skin color, muscles, hips, and breasts.

Despite its fascination with certain Hellenic models, Jay (1993, 269) notes, Heidegger's thought can be interpreted as "recovering the Hebraic emphasis on hearing God's word rather than seeing His manifestations." In this sense, Heidegger—specifically in "The Age of the World Picture"— depicts certain consequences of Noah's curse in the technological transformation of twentieth-century life. The concept of the "world picture," Dalia Judovitz (1993, 84) points out, does not have a "pictorial referent, but rather refers to the manner in which the world is conceived, that is, represented (both composed and apprehended)." For Stephen Houlgate (1993, 91), the concept of the "world picture" denotes an age in which we conceive of whatever is as something "placed before us," as objective. To his critics, Heidegger succumbed to this conception of the world during the Nazi period (see Morris and Weaver 2002; Pinar 2002c; Jay 1993, 269).

It is now widely known that Foucault was focused on the social and political consequences of ocularcentrism, among them modern technology, and modern forms of "governmentality," focused in the image of the panopticon.

For Foucault, the Enlightenment had been betrayed by the panopticism of its technologies, each of which—of power, of self—pointed toward processes of totalization, normalization, and domination (Levin 1993b). For Foucault disciplinary power—such as whiteness—operates through its invisibility. The objects of discipline, Mason Stokes notes, are subjected to a regimen of compulsory visibility. Stokes quotes Foucault to make his point: "It is the fact of being constantly seen, of being able always to be seen, that maintains the disciplined individual in his subjection" (1975, 187; quoted in Stokes 2001, 160). That was how "power" played inside the tent: Ham disciplined his father by looking at him, disciplined him as his sexual servant. Noah retaliated, relegating Ham's progeny to specularized servitude and economic exploitation.

Why? The fantasy of panopticism, which is a specular fantasy of omniscience, follows, I suggest, from the denial of lack, the mythic movement of which was the "curse of Ham." While such gendered specularity is ancient, the hegemony of vision at work in modernity is apparently historically distinctive, in part because it is allied with and embedded in contemporary technologies. For Levin (1993a, 7), panopticism is the "political" performance of "enframing," the deployment of technologies of "control." It is during modernity, he suggests, that the ocularcentrism of the West takes the political form of panopticism. During modernity administrative institutions are established to enforce those disciplinary practices made possible by the "conjunction" of a "universalized rationality" and "advanced technologies," practices designed to secure the "conditions" of "visibility" (Levin 1993a, 7). Through observation, "truth" becomes observable, measurable, scientific: order is maintained (see Rogoff 1996, 189).

Levin (see 1993a, 14) suggests that Heidegger was (in contrast to Derrida) neither against the *hegemony* of vision as such, nor even (in contrast to Descartes and Sartre) critical of vision; nor did he regard listening as an end in itself (see Silverman 2000, 69). Rather, Levin continues, Heidegger's thinking was inspired by the "ocularcentric Greeks," indebted to their vision-generated, vision-centered language. It was, Levin asserts, the *character* of the vision that triumphed in Western ocularcentric culture and in its philosophical discourse that so deeply troubled Heidegger. It was this distinctive character of ocularcentrism that he held responsible for the increasing "darkness" of the world, our increasing "closure" to the light or unconcealment of being (Levin 1993a, 14). This "pathological" vision constituted what Heidegger characterized as the age-of-the world picture, an age when being itself is reduced to enframing of representation (see also, Mitchell 1994).

Other philosophers and critics of the age have focused on visual culture (see, for instance, Rogoff 1996, 189–190). Levin (see 1993a, 23) points to Walter Benjamin's identification of the distinctiveness of modernity as its visual productivity and visual obsessiveness. For Benjamin, the age is structured by dream-images and commodified visual fetishes, in Levin (1993a, 23) words, "visual processes re-enchanting" the world which the

"Enlightenment," and then "Marxism," had "struggled" to "free" from "illusion." These visual processes, he continues, mask the "violence" of the social world. Sounding like Avital Ronell (1992), Levin (1993a, 25) asserts that our intensifying "seduction" and "narcotization" by visual images enables the market economy to replace the public square with the shopping mall. Public debate and dialogical encounter disappear, as "communication itself" is replaced by "visual narrative." This point Levin (1993a, 24) underlines by quoting Michel de Certeau: "From television to newspapers, from advertising to all sorts of mercantile epiphanies, our society is characterized by a cancerous growth of vision, measuring everything by its ability to show or be shown, and transmuting communication into a visual journey." Robyn Wiegman's "economies" of "racial visibility" come sharply into focus: ongoing debates over civil rights and racial justice are replaced by interracial buddy films in which, once again, black men serve the needs of white.

Economies of Visibility

Whiteness and masculinity are . . . complex cultural fantasies. (Thomas DiPiero 2002, 230)

[T]he inscription of Otherness [is] on the black body. (Lola Young 1996, 50)

[O]verdetermined black materiality is indispensable to the production of white social transparency. (Russ Castronovo 2001, 183)

What are black men in the American ocular imagination? (Elizabeth Alexander 1996, 160)

Robyn Wiegman (1995, 195) proposes a racialized "economy of visibility," wherein the white "critical gaze" is "trapped" within its own "seeing" (1995, 2). In such an economy, the relations of "sight" and "observation" associated with the Greek *theoria* (defined as "a looking at") are intertwined with contemporary forms of modern disciplinarity (in which "surveillance" figures), expressed through a "spectatorial subjectivity" (De Bolla 1996, 68). The association between theoretical investigation and the primacy of sight is, Wiegman (1995, 3) suggests, a "cornerstone" of modernity, more specifically, one of its most "anxious" and "contentious epistemological productions." Acknowledging that vision is the privileged sense of modernity, Wiegman points out that its epistemological status as the means by which both meaning and truth are established is undermined, not only by "theory," but by shifts in technological production and reproduction as well.[2]

Despite the scholarly criticism of the hegemony of visuality, Wiegman (1995, 3) points out that the "popular realm" of the "visual" constitutes, in our time, a "newly configured public sphere." What can it mean, Wiegman wonders, that photography, film, television, and video (as well computer technologies) function as our primary public domain, the shared space in which cultural politics occurs? What can it mean that within the visual-popular sphere the body becomes the primary site of representation? These questions carve the context in which Wiegman (1995, 3) reads certain "convergences" and "divergences" between gender and race. Although these "economies of visibility" (1995, 3) are clearly associated with their development, their reach is hardly restricted to that concept.

Wiegman points out that the production of the black African body as non- or subhuman, as object and property, cannot be adequately understood as the ideological rationalization of the slave trade. What provided the cultural precondition for seeing "blackness" as "subhuman" and as "property"? Significantly, Wiegman (1995, 4) points to the "binary" structure of vision. Visuality was both an economic system and a representational economy, thereby naturalizing racial categories as "real," observable and, later, measurable, as late-nineteenth-century preoccupations with skin, hair, breast, brain size, and skull shape illustrate (see Stoler 1995). Nor was it historical coincidence, Siobhan Somerville (2000, 3) points out, that the "classification of bodies as either 'homosexual' or 'heterosexual' emerged at the same time that the United States was aggressively constructing and policing the boundary between 'black' and 'white' bodies."

It was a racializing and engendering visuality in the service of narcissism, Lee Edelman suggests. In his discussion of Fanon's *Black Skin, White Masks,* Edelman (1994b, 46) points out that in conjoining blackness and whiteness— "the black . . . must be black in relation to the white man," Fanon (1967, 110) appreciates—the narcissism embedded in the mythology of "racial" supremacy is "re-enacting" the "logic" of "phallic masculinity under compulsory heterosexuality." This is, Edelman (1994b, 46) explains (quoting Fanon 1967, 110), a logic of "visual difference" requiring the "display of the 'other' in the position of 'lack' in order to reassure the dominant subject, by contrast, of his (phallic) 'possession'," Edelman (1994b, 46) continues:

> This fantasy of "possession" comes to signify, in turn, with the facility of a metonymic slippage, possession of the (fetishized) gaze that objectifies and commodifies the other-as-(b)lack, thereby effecting, in the historical context of colonization and enslavement, the racist interpretation of the black body as material supplement, as that which is to-be-possessed.

Edelman understands the curse of Ham exactly.

These "economies of visibility" produced networks of racialized and gendered meanings attached to bodies; as such, Wiegman asserts, they were not only political practices: they were and remain today subjective as well. It is precisely this convergence of the political with the subjective that has given both power and substance to identity politics in the last two hundred years (Wiegman 1995). As V.Y. Mudimbe (1988, 12) has observed: "The African has become not only the Other who is everyone else except me, but rather the key which, in its abnormal differences, specifies the identity of the Same." The twentieth-century fascination with Tarzan illustrates Mudimbe's final phrase (see, for instance, Bederman 1995).

Despite the binary structure of vision—whiteness was, presumably, whatever blackness was not, even when whiteness copied then bleached blackness, as in the case of the Tarzan fantasy—Wiegman points out that human bodies are neither black nor white, and that the binary contradicts what in fact the white eye observes. It is perhaps this contradictory character of racial "observation" that has enabled the slippage from "race" to gender and back again. Both slaveholders and abolitionists emphasized the gendered aspects of enslavement. The African's ontological status can be read, Wiegman (1995, 11) suggests, as "hinging, in part, on sexual difference." In the case of white men's gendering of black men, it would seem to be a case of self-same sexual difference. From gender, race appeared.

Within the Western cultural imaginary, did European and European-American men subliminally construe black men as "Adam's rib"? Self-difference disavowed becomes alterity. In aligning black men with women, Wiegman (1995, 14) notes, white men positioned black men in a "structurally passive" or "literally castrated" sphere of "sexual objectification" and "denigration." This is a succinct statement of the queer character of race in America. These white repositionings of the black male were compensatory and defensive denials of an ongoing "crisis" of white masculinity (see Pinar 2001, chapters 6 and 19, for an account of the U.S. version). Moreover, while attempting to castrate the black man in a civic as well as literal sense, white men were also acknowledging that they and black men were, in white men's minds, sexually associated. Both women and black men were relegated to the sphere of "sexual objectification" and "denigration."

When late-nineteenth-century white men thought of (white) women, they "saw" black men threatening to rape them. "For what the eye sees," Wiegman (1995, 24) points out, is itself a "complicated" and "historically contingent production." In the initial sighting of the African's "blackness," Wiegman underscores, was the economic and epistemological genesis of enslavement, as well as those subsequent "scientific" schemes of "difference" organized around cranial capacity and anatomical form. Was the genesis of enslavement a traumatic restructuring of sexualized anatomical sameness? Are racial hierarchies also a defensive resymbolization of (white male) self-shattering at the (sublime) sight of the black male body?

Wiegman turns not to the literary queer theory of Leo Bersani (1995) or the psychoanalytic theory of Kaja Silverman (1992) but to the historical theory of Michel Foucault. There is the argument that in the seventeenth century, the "classical" mind—devoted to discovering truth by "drawing things together"—gave way to the modern mind, devoted "to proof by comparison" (Foucault 1973, 55; quoted in Wiegman 1995, 25). Such a comparison pivoted on "the apparent simplicity of a description of the visible," and in this, Wiegman notes, natural history's epistemological reliance on rationalistic vision began. Visuality, she notes, was no longer linked to the other senses: the eye was enjoined "to see and only to see" (1973, 43; quoted in Wiegman 1995, 25). As in whiteness, subjectivity becomes invisible in observation and measurement.

The rationalization of vision and comparison as a structure of thinking produced two "primary" and "exclusionary figures," namely "identity" and "difference" (Wiegman 1995, 25–26), "restricted . . . to black and white" (Foucault 1973, 133; quoted in Wiegman 1995, 26). It is within this historical and epistemological context, with the observer's neutrality apparently established by the methodology of observation and comparison, that natural history's increasing interest in human classification becomes intelligible. It is in this context, Wiegman (1995, 26) points out, that the "rendering" of race as an "epiphenomenon" of the skin was "damagingly drawn." Although the systematization of difference as binarized into black and white characterizes modernity, the originary fantasy of difference is, of course, premodern, indeed, Edenic. It is between God-the-Father and Man-the-Son, then between Man and (Wo)Man.

The fetishization of the black body did not, of course, stop at the skin. The seventeenth-century observer Richard Johnson "reported" that blacks are "furnish[ed] with such [sexual] members as are after a sort burdensome unto them," and the late eighteenth century English surgeon Dr. Charles White "observed" "that the PENIS of an African is larger than that of an European has, I believe, been shewn in every anatomical school in London. Preparations of them are preserved in most anatomical museums; and I have one in mine' " (quoted in Hoch 1979, 52). Parts of lynching victims would sometimes be stored in jars so the white Southern public could look at them (see Pinar 2001).

The historical emergence of visuality's dominance was not limited to natural history, of course. Indeed, its systematical structuring of vision and form was, Wiegman (see 1995, 26) notes, a version of Cartesian perspectivalism. Wiegman alludes to Martin Jay's elaboration of a relationship between the English Renaissance's aesthetic understanding of vision and the scientific perspective founded on the philosopher's dispassionate eye, a relationship that structures, for Wiegman (1995, 26), the broad historical context in which the African's "blackness" became visible.

Linked to the "objective optical order" heralded by the artistic theory and practice of the Italian Quattrocento, Cartesian perspectivalism was characterized as "a lone eye looking through a peephole at the scene in front of it . . . static, unblinking, and fixated" (Jay 1988b, 6, 7; quoted in Wiegman 1995, 26). Nicely summarizing both Jay and Foucault, Wiegman (1995, 26) points out that this "observing eye" and its ordering relation to a hierarchical exterior express the quintessential scene of observation wherein visuality constituted itself epistemologically, by means of "emotional withdrawal" and "bodily repression," at the "radiating edge" of the "monocular eye." The sexually saturated look in Genesis 9:23 becomes bleached in the modernist compulsion to observe and measure.

Not only the natural world was classified and "disciplined," of course. The social world is organized and disciplined as well, as readers familiar with Foucault well know. Wiegman quotes Foucault's characterization of sixteenth- and seventeenth-century penal practices, specifically the disciplinary power of spectacle: the appearance of the "tortured, dismembered, amputated body, symbolically branded on face or shoulder, exposed alive or dead to public view" (1975, 8; quoted in Wiegman 1995, 37). In the eighteenth century, spectacle becomes systematized as panopticism and surveillance. As Foucault's concepts make clear, the structuring of the visible was central to discipline and punishment. The taboo of looking resurfaces as the racialized refusal to be seen. As Mason Stokes (2001, 161) points out, the subtitle of the Ku Klux Klan, "The Invisible Empire," points to the racialized "anxiety over being seen and recognized."

This anxiety is also evident in Frederick Douglass's 1853 "The Heroic Slave," in which Douglass forefronts black subjectivity in his description of the black male body. This first known short story published by an African-American man illustrates, Wiegman (1995, 71) suggests, not how the slave's search for subjectivity is embedded within textual economies of both race and gender, specifically in "narratives" of "masculinity." Douglass' description of the black body, Wiegman (1995, 74) continues, "adjudicates" the slave's "specular particularity" through aesthetic language evoking a familiar masculine "ideal."

> Madison was of manly form. Tall, symmetrical, round, and strong. . . . His torn sleeves disclosed arms like polished iron. . . . His whole appearance betokened Herculean strength; yet there was nothing savage or forbidding in his aspect. . . . A giant's strength, but not a giant's heart was in him. His broad mouth and nose spoke only of good nature and kindness. . . . He was . . . intelligent and brave. He had the head to conceive, and the hand to execute. (1972, 40; quoted in Wiegman 1995, 74)

As Wiegman points out, this description reiterates nineteenth-century preoccupations with antiquity and, particularly, with the presumed resonance between male physical beauty and character (Mosse 1996). But it also demonstrates the "impossibility" of "uncovering" a black presence within the "universalized particularity" of white masculinity (Wiegman 1995, 75). That sleight-of-hand in which white particularity, and, specifically, male embodiment, disappears into universality, accompanies the racialized disavowal of homosexual desire.

Wiegman (1995, 75) decodes the passage as "displacing" race by "sexual difference," thereby enabling Douglass to invoke the slave's masculinity as the common—and civic—bond he shares with white men. Wiegman does not fail to notice the absence of African-American women in this formula. "Women" disappear in their conjunctive relationship with black men, positioned there as white men struggled to disentangle themselves from the "homosocial nexus" of "violence" and "heightened erotic passion" characterizing the late-nineteenth-century scene of lynching and castration (Wiegman 1995, 77). "Race" becomes "displaced" by gender because, in the white mind, "race" has always been a recoding of gender, a displacement of disavowed self-same sexual difference onto the abjected, now racialized, other. In such a homosocial system of desire, *la femme n'existe pas.*

CHAPTER 4

Outside the Tent

A New Covenant

Christianity, a religion founded upon a wounded god, particularly valorized wounded bodies. (Gary Taylor 2002, 39)

The seed is the word of God. (Luke 8:11)

Although I saw with the black vision of Ham, I was, I suppose, as pious as Shem and Japheth. (Ralph Ellison 1995 [1964], 117)

[T]he knowledge of the object cause of the master's own desire is denied him. (Saul Newman 2004, 306)

One may be subjugated but saved, Jesus promised. (Or is it through subjugation that one is saved?) In accordance with Jewish law, Jesus was circumcised on the eighth day (Luke 2:21). However, the practice was not exactly prominent in his teachings. It was only after his death that the question of circumcision arose when Jesus' apostle Paul (a Jew steeped in rabbinic tradition) began to proselytize Gentiles. Of course, the first male Christians were Jews who had been circumcised; many of these early church members favored compulsory circumcision for converts. But Paul (whom Gollaher [2000, 31] characterizes as a "genius" of "practical evangelism") realized that requiring circumcision would reduce the number of eligible converts. Given the pain and suffering that accompanies the operation, few men would have considered Christianity had circumcision been a prerequisite (Gollaher 2000).

Quite the strategist, Paul reinterpreted the ancient distinction between physical and spiritual circumcision. In his Letter to the Galatians, Paul suggested that in Christ a new covenant between God and humankind had been struck, one that subsumed the old covenant between God and Abraham. Christ, Paul declared, fulfilled the law, and this fulfillment rendered the ancient covenant, and its key marker, circumcision, passé (Gollaher 2000). "In Christ Jesus neither circumcision nor uncircumcision counts for anything," Paul proclaimed (Galatians 5:6). In Corinth, where conservative Jewish converts were pressuring their Gentile counterparts to become circumcised, Paul was insistent: "Was anyone already circumcised when he was called?

Let him not seek to remove the marks of circumcision. Was anyone uncircumcised when he was called? Let him not seek circumcision" (1 Corinthians 7:18).

Faith in Christ eliminated the point of circumcision, Paul argued. No longer was it necessary to distinguish Jews from Gentiles. Unlike the law of the patriarchs, the new covenant was universal. Determined to preach a simple Christian faith that transcended the elaborate, highly codified law with which he had come of age, Paul pointed to circumcision as the marker, not of the covenant with the Father, but of an old, outmoded order (Gollaher 2000). Thus in his Letter to the Romans he criticized what he considered Jewish legalism:

> Thou that makest thy boast of the law, through breaking the law dishonourest thou God? For the name of God is blasphemed among the Gentiles through you, as it is written. For circumcision verily profiteth, if thou keep the law: but if thou be a breaker of the law, thy circumcision is made uncircumcision. (Romans 2:23–25)

Gollaher points out that this passage is from the traditional King James version of the New Testament, a version often euphemistic in its treatment of the sexual. He reports that a more accurate translation reads: "If you break the Law, your circumcised glans becomes a foreskin" (quoted in Gollaher 2000, 33).

This was a point Paul would make over and over: failure to live up to any part of the old covenant made a Jew a sinner, no different in God's eyes than an uncircumcised Gentile. Those, including Gentiles, who, through faith in Christ, accepted the "righteousness of the law" were, equally with Jews, the legitimate heirs to the covenant of Abraham (Gollaher 2000). Circumcision was, then, irrelevant to the new covenant.

> For he is not a Jew, which is one outwardly; neither is that circumcision, which is outward in the flesh. But he is a Jew, which is one inwardly; and circumcision is that of the heart, in the spirit, and not in the letter; whose praise is not of men, but of God. (Romans 2:28–29)

In the new covenant, the mark of the father was to be made, not on the penis, but in the heart.

Be that as it may, Paul knew that God's covenant with Abraham was marked by physical circumcision. How could he accommodate that fact with his argument? The following discursive move seemed to suffice: Paul insisted that, according to a close reading of the book of Genesis, "faith was reckoned to Abraham for righteousness" *before* he was circumcised (quoted in Gollaher 2000, 33).

> And he received the sign of circumcision, a seal of righteousness of the faith which had yet being uncircumcised: that he might be the father of all them that believe, though they be not circumcised; that righteousness might be imputed unto them also. And the father of circumcision to them who are not of the

circumcision only, but who also walk in the steps of that faith of our father Abraham, which he had being yet uncircumcised. For the promise, that he should be the heir of the world, was not to Abraham, or to his seed, through the law, but through the righteousness of faith. (Romans 4:9–13)

Many early Christians preferred not to take any chances. Not only did circumcision persist, so did self-castration, a practice sufficiently widespread that it had to be banned in 395 by Pope Leon I (see Taylor 2002, 73).

In Ephesians, an epistle written a generation or so after Paul's death that sought to develop his theological formulations more fully, the writer reiterates the same point, namely that while the newly converted are now heirs to God's covenant, they were once "Gentiles in the flesh, those who are called 'foreskin' by those who are called 'circumcision' " (Ephesians 2:11). Conscious of the theological significance of the blood shed during *berit milah*, the writer suggests that Gentiles, by accepting Christ's blood sacrifice on the cross, are "vicariously circumcised" (Gollaher 2000, 33).

The early Church's suspension of the practice was, Gollaher (2000, 34) suggests, a "crucial" feature in Christianity's conversion from a Jewish sect to a community with a recognizably distinct religious identity. Among these early Christians there was often the sense that Jesus had freed them from all ritual law, from circumcision to dietary restrictions. "We Christians eat pork," a Christian boasted in the seventh-century Trophies of Damascus, "because He who freed me from circumcision also freed me from abstinence from pork" (quoted in Gollaher 2000, 34). Was he freed as well from the Leviticus condemnation of homosexuality? Paradoxically, Christians still accepted the Torah and other books of the "old" testament as the literal word of God (Gollaher 2000).

The early Alexandrian theologian Origen evidently considered circumcision an insufficient gesture; like many early Christians, he castrated himself (see Taylor 2002, 39). Like Origen, circumcision was insufficient for Peter Abelard, the twelfth-century French monk and theologian, another Christian theologian who struggled over the significance of his genitals in his relationship to the Father. Recalling Matthew 19:12, where Jesus acknowledges those who "have made themselves eunuchs for the kingdom of heaven's sake," Abelard believed that the mutilation of his genitals had been ordained by God:

> The hand of the Lord had touched me for the express purpose of freeing me from the temptations of the flesh and the distractions of the world so that I could devote myself to learning, and thereby prove myself a true philosopher not of the world but of God. (Quoted in Taylor 2002, 40)

Where did the Father touch him? Was castration—was circumcision—compliance with or resistance to Jesus' injunction (concluding Mathew 19:12) "let him, that is able to receive it, receive it"?

The self-mutilation Abelard imagined the Father wanted him to perform on himself was not recommended, however, to others. Indeed, Abelard thought even circumcision was optional. After all, Enoch, Noah, and Job, not to mention Moses, had presumably entered the kingdom without cutting the penis. Abelard argued that circumcision had been intended for Abraham and his male offspring only. Since circumcision had not, in fact, served as a reliable sign of salvation and was explicitly rejected by the apostle Paul, the philosopher in Abelard's dialogue rejects it as obsolete, along with other eso-teric temple rituals specified in the book of Leviticus (Gollaher 2000).

This hardly settled the matter, as Abelard continued his argument in two other works, *Commentaries on Romans* and *Sermon on Circumcision*. There had been a time, he allowed, when circumcision had served a purpose. The mark had set Israel apart from the Gentile tribes around them; it facilitated the marriage of Jews to other Jews. The coming of Christ, Abelard reasoned, dissolved the need for any distinctions. "With the cessation of the Law and the succession of the more perfect Gospels," he wrote, "circumcision has been overtaken by the sacrament of baptism which sanctifies men and women alike." Other theologians agreed. Citing Paul's statement that "in Christ there is neither male nor female," Guibert of Nogent concluded that it was unthinkable that any measure of saving grace would exclude women (quoted passages in Gollaher 2000, 34).

Peter Alfonsi reiterated Paul's pronouncement that in Christ, the distinc-tion between Jew and Gentile vanished. Judaism, Rupert of Deutz observed, was tribal. Mixed among dozens of other desert peoples, the Jews evidently needed a distinguishing characteristic. (This hardly solves the mystery why the mutilation of the penis was selected as the distinguishing characteristic.) Christianity was not bounded by tribal identity: its promise, presumably, was salvation to all peoples, of all languages and "tribes." Certainly less painful than circumcision and more easily applied to the multitudes, baptism had also been, in the teachings of Christ and the apostles, directly linked to salvation. For Christians, then, circumcision symbolized a practice no longer necessary after Christ (Gollaher 2000).

The most "rigorous" effort, David Gollaher (2000, 35) judges, to posi-tion circumcision in Christian theology was made in the thirteenth century by St. Thomas Aquinas. In his *Summa Theologica*, Aquinas attended to the spiritual significance of circumcision. Many of his colleagues, however, were focused on a seemingly more concrete topic: they pursued a serious debate over the ontological status of the foreskin of Christ. Faith in the resurrection of the body and images of Christ as a corporeal being sitting at the right hand of God led to the question of whether, after the Ascension to heaven, Christ had recovered his foreskin. Finally, theologians realized that this question could not be answered (Gollaher 2000).

Before and during the Renaissance, one of the most prized relics was the foreskin of Christ. Gollaher reports one legend in which Mary saved her son's prepuce and carried it with her until she ascended to heaven. There she pre-sented it to Christ so that "he might stand intact before God-the-Father" (Gollaher 2000, 36). This seems quite the conventional oedipal fantasy: the mother restores the "wholeness" of the son so he might "face" his father. Other legends, Gollaher notes, leave it on earth to be discovered later by church fathers. In one story Mary, the mother of Jesus, gave the foreskin to Mary Magdalene who, before her death, gave it to the apostles. In the *Revelations of Saint Birgitta*, a Swedish saint (who was canonized toward the end of the fourteenth century) claimed that Mary appeared to her in a dream, telling her that she had herself preserved the blessed foreskin and, finally, passed it onto the disciple John (Gollaher 2000). Given the fantasy of the Virgin Mary—Dyer (1997, 74) designates her the "supreme exemplar" of "feminine whiteness" in that she reproduces without sex—her interest in her son's foreskin seems strange indeed.

In another legend, the foreskin survived until the time of Charles the Great in the late eighth century when an angelic courier, in anticipation of Charlemagne's coronation by Pope Leo XIII in the year 800, brought the divine relic to him. The emperor presented the foreskin to the church where it remained a private possession of the popes until the invasion of Rome in 1527. Then, one of the soldiers of Charles V stole the foreskin, setting the scene for the miracle of its recovery. This legend also held that the foreskin emitted a "sublime odor," delighting the "grand ladies" of Rome (Gollaher 2000, 37). Its return to Rome was pronounced a miracle (Gollaher 2000).

Clearly, eroticism was a common undercurrent in medieval and Renaissance fantasies of Christ's foreskin. In the mystical vision of the Austrian saint Agnes Blannbekin (d. 1315), sexual themes are explicit. In her *Via et Revelationes* she described a vision in which she swallowed the divine foreskin: "She feels a small membrane on her tongue, like the membrane of an egg, full of exquisite sweetness." When she touched the membrane with her finger, it slid down her throat: "so great was the sweetness at the swallowing of the membrane that she sensed a sweet transmutation through the muscles and organs of her whole body" (quoted passages in Gollaher 2000, 37). Afterward, she claimed to have been able to re-experience this divine orgasm whenever she touched her tongue with her finger. St. Agnes's experience of the Eucharist as an erotic event, supplemented by another vision in which she described seeing Jesus nude in a river, were judged by her clerical superior as pornographic, and *Via et Revelationes* was long suppressed (Gollaher 2000). Is the Eucharist, for women, a ritual symbol of—if not fellatio proper—swallowing the foreskin of Christ?

A THEORETICALLY INDISPENSABLE SCENE

[T]he very symbolic order that is sustained through homosexual desire prohibits the direct expression of that desire, obliging it to assume the censored and circuitous form of heterosexual exchange. (Kaja Silverman 1988, 185)

[T]he homophobia of the church . . . is a defense against its own homosexual latency. (Tom F. Driver 1996, 54)

[T]he son's identification with the father represents the introjection of his desire for him. (Mark Simpson 1994, 75)

The prohibition against male–male sexual acts appears in a list of sexual and other offenses, including various kinds of incest, bestiality, intercourse with women who are menstruating, and dedicating one's child to Molech. It is not clear, Eilberg-Schwartz acknowledges, what unites this list, although, he notes, many of these acts pose a threat to the integrity of Israelite lineage or waste Israelite "seed." In Genesis 38:8–10, when Onan "spilled his seed upon the ground," the Lord was so displeased that he struck him dead. Thomas Laqueur (2003, 112) points out that if it was masturbation that was to blame, the event would be the "true ground zero of solitary sex"—but, in fact, we do not know how Onan managed to make this fatal error.

Gary David Comstock (1991) reminds us that the prohibitions against and prescription of death for male–male sexual acts consists of two verses, a small number relative to the entire volume of scripture and small even in the context of the book of Leviticus itself (859 verses). He notes that even the lists of sexual offenses, to which these verses belong, represent neither a large part nor a major theme in Leviticus. Comstock also points out that the punishment of death is hardly limited in Leviticus to homosexuality. Other such punishable acts include cursing father or mother, committing adultery with neighbor's wife (both for man and woman), committing incest with the father's wife, committing incest with the son's wife, marrying a woman and her mother, committing bestiality, being a medium or wizard, being a harlot (if the daughter of a priest), working on the Sabbath, cursing or blaspheming the name of Yahweh, and killing a person (Leviticus 20:9–16, 20:27, 21:9, 23:13, 23:16, 23:30, 24:17, 24:21).

"One looks in vain for an example of inclusive community, egalitarian principles, or a theology of loving outreach and pluralistic justice in Leviticus," Comstock (1991, 126) laments. "Leviticus is," he continues, "about defining a separate community that sets itself apart by virtue of its superior differences with others." He notes that other, nonsexual regulations, specifically the social legislation, appear to function similarly, that is, underline ancient Israelite exclusivity by mandating mutual support among those who are the "insiders."

Leviticus is a "patriarchal document," Comstock (1991, 132) observes. He notes that the sexual regulations in chapter 18 are addressed to men and forbid sex with one's mother, aunt, step-mother, aunt by marriage, sister, half-sister, step-sister, step-daughter, sister-in-law, daughter-in-law, grand-daughter, step-grand-daughter, and neighbor's wife. Conspicuously absent from the list are daughters (or sons), a fact, he reports, biblical commentators explain away by suggesting that these were assumed as forbidden or omitted by accident. However, he continues, feminist scholars have pointed out that the sexual violations are phrased not as offenses against women but as offenses against men who violate rights of ownership, use, and exchange. A patriarch owned his wife, and he owned his daughter until he gave her away in marriage; even then, as we noted in the Sodom story, it was possible to give daughters away even after their engagement, even to a riotous mob of would-be rapists. By prescribing capital punishment for sex only with the neighbor's wife, father's wife, and daughter-in-law, chapter 20 makes it clear that ownership, not consanguinity, is the key concern in these sexual regulations. "It is the sexual use of those women who belong to other male relatives or fellow-countrymen that is forbidden," Comstock (1991, 132) concludes.

"Because the prohibitions against male homosexuality and the prescription of death for it follow immediately the preceding regulations in both chapters 18 and 20," Comstock (1991, 132) continues, "it would seem that for a man not to possess a woman sexually, to possess a man as a woman, or to allow oneself to be possessed as a woman are also extremely serious violations of patriarchal behavior." In Leviticus, Comstock adds, sexual violations are delineated as they are nowhere else in Hebrew scripture (cf. Exodus 20 and 21; Deuteronomy 27). Elsewhere death is not prescribed for sexual violations; male–male sexual acts are not even mentioned. "Relative to other sets of laws in Hebrew scripture," he argues, "these lists appear as desperate attempts to delineate, exaggerate, and apply patriarchal principles" (1991, 133).

Howard Eilberg-Schwartz, too, points to the exclusivity function of the Leviticus statutes, noting that the list of prohibitions begins and ends with the injunction not to imitate the ways of the Egyptians or the Canaanites, both groups imagined to be the descendants of Ham (Gen. 10:6). He suggests that in ancient the Israelite imagination, male–male sexual acts were disavowed as alien and, consequently, thematized their stereotyping of its proximate "others," the Canaanites. The association of male–male sex with Canaanite social practices in the Holiness Code supports, Eilberg-Schwartz argues, his interpretation of the Noah story. In Noah's cursing of Canaan for his father's "sin," the Canaanites became an immoral people. The ancient Israelites employed the same "genealogical" strategy to defame Moabites and Ammonites, declared to be descended from the incestuous union of Lot and his daughters (Gen. 19:30–38), which, Eilberg-Schwartz notes, repeats in significant ways the story of Noah and his son Ham. These offspring of incest were, presumably, the ancestors of the Moabites and Ammonites. Israel, Eilberg-Schwartz summarizes, emerges (in its own mind) as one of the few genealogical lines untainted by sexual perversion (see, also, Schwartz 1997, 107).

What can be the "wound," Lee Edelman (1994a, 266) asks, the scene of sodomy "inflicts," so that its "staging," even imaginary, seems so "dangerous?" It is, Edelman (1994, 266) asserts, a "scandal" of "supposition" implicit in the psychoanalytic articulation of the constitution of masculine subjectivity, centered on the "crisis of representation" through which the subject comes to imagine sexual difference. Working from Freud's *From the History of an Infantile Neurosis*, Edelman argues that the male subject's construction of sexual difference follows from issues of castration and a "retroactive understanding" (1994a, 266) of what Freud construes as the primal scene, a scene, it turns out, of sodomy, a scene from which, Edelman suggests, Freud himself flees.

First Freud will assume the position of the one who sees, which is to say, the one who knows. In imagining the primal scene (a scene of identificatory uncertainty and diffused division between real and imagined, external and internal, patient and psychoanalyst), Freud is, Edelman argues, substituting himself for the patient's father. The psychoanalyst's imaginative reconstruction of the scene, Edelman (see 1994a, 273) points out, can thus be decoded as a reenactment of the scene, as the analyst is now the one who imposes himself, not unlike the unconscious, as if to escape the patient's notice. Freud stands accused, then, of using his patient from "behind." It is Freud's own ambivalence, Edelman argues, over this scene of penetration from "behind" that stimulates Freud's own defensiveness about the status of his own analytic interpretation.

This defensiveness Edelman ascribes less to Freud's own sexual ambivalence (that seems well established, specifically in his relationship to Fleiss), but rather to the threat suggested by the sodomitical scene itself (see Edelman, 1994a, 271). To elaborate this point, Edelman (see 1994a, 273) quotes Stanley Fish: "Freud reserves to himself . . . the pleasure of total mastery. It is a pleasure that is intensely erotic . . . affording the multiple satisfactions of domination, penetration, and engulfment." Although accepting Fish's characterization, Edelman (1994a, 273) suggests that neither mastery nor pleasure adequately describes Freud's response. Rather, Edelman (see 1994a, 273) proposes that the instability of Freud's interpretative position follows from the very witnessing of the primal scene itself. It is a scene, Edelman (see 1994a, 273) emphasizes, that inevitably wounds the spectator.

As Freud himself understood, there is no defense against those critics who insist on reading Freud's interpretation of the primal scene as *his* fantasy, disclosing *his* desire. "On the one side," Freud acknowledged, "there will be a charge of subtle self-deception, and on the other of obtuseness of judgement; it will be impossible to arrive at a decision" (quoted in Edelman, 1994a, 273). The primal scene *is*, it would seem, radical indeterminacy, in part due its destabilization of the therapist, now threatened to be repositioned as the patient (see Edelman 1994a, 274).

Edelman discerns such interdeterminacy in Freud's final remark concerning the status of the scene itself: "I intend on this occasion," he declares, in a passage added after he had finished the manuscript of his text, "to close the discussion of the reality of the primal scene with a *non liquet*" (quoted in Edelman 1994a, 274). This is, Edelman points out, a legal judgement that the evidence in the case is insufficient. But the Latin phrase made over by the law means literally, Edelman notes, that "it is not clear," a return of the optical metaphor always at issue in any construal of the primal scene. By acknowledging the absence of clarity in his ability to view or "catch a glimpse" of what for Freud has become, as Edelman (1994a, 274) points out, a "theoretically indispensable scene," Freud locates himself ambiguously before the very *analytic* scene from which the Wolf-Man's primal scene was formulated. Freud's acknowledgment, Edelman (1994a, 274) continues, of a "conceptual opacity" within psychoanalysis "reenacts" the resistance of the Wolf-Man to his spectatorial engagement in the primal scene. What Freud no longer wants to "see" is his implication in the primal scene *a tergo*, his substitution of himself for the Wolf-Man's father, the father the Wolf-Man wants, the father the Wolf-Man imagines as seductive.

In being persuaded that Freud's construction of his primal scene is accurate (even though he never remembers it), the Wolf-Man is allowing himself to be seduced by his father-analyst. Freud knows that the Wolf-Man's sexual formation derived from his identification with the pleasure associated with the penetration of the anus, a "penetration" interpreted as both the act of penetrating and the act of being penetrated (Edelman 1994a). Curiously, this same ambiguity haunts Genesis 9:23.

That passage tells us that "Noah knew what his son had done unto him." If what Ham had "done" was explicitly (rather than implicitly, as in gazing at the naked body of the patriarch) sexual, then it would appear, at first blush, that what he had done was to penetrate the patriarch. The passive tense of the verb "to do" suggests that. (Whether the translation is accurate is secondary to the meaning the translation conveys to English readers. I am suggesting what the passage might mean given the translation, and am less interested in whether or not the translation coincides with earlier denotations.) But "suggests" is all it does. It could mean that the son, as had Lott's daughters after the destruction of Sodom, seduced him, made his father enter him, making his son into a "daughter," destroying the manly character of Noah's progeny and legacy. The son so feminized is, by definition, "cast out," no longer his "son," and his "castrated" status condemns all who come from his seed. To be a slave in this sense suggests being without patriarchy, without a father and without progeny (i.e. a de-generate), relinquishing the patriarchal status that being a husband and father should confer (and did confer in ancient Israelite culture). It also threatens the sublimated position of the patriarch who becomes, to borrow a contemporary example, the priest who abuses the altar boy, a man of God whose "honor" has been desublimated by the pleasures of the flesh.

The Wolf-Man's double identification—as the one penetrated and the one penetrating—allowed him to occupy the positions of both his mother and his father in the primal scene. As an adult, the Wolf-Man reproduced this primal scene in his sexual practices, for example, the anal penetration of women and his own penetration by men administering enemas. His guilt was expressed by the wolves, a dream Freud interpreted as a moment of castration. It was, Freud imagined, castration that was the price paid for gratifying his patient's "homosexual enthusiasm" (quoted in Edelman 1994a, 274). It would seem Freud is making here a Foucaultian disciplinary move by assigning sexual identity to a complex and multiply focused desire.

Edelman notes that the Wolf-Man's dream not only recodes the original—primal—experience of sexual diffusion and ambivalence, it does so by reassigning spectatorial and thereby identificatory positions. The boy who viewed the primal scene (and in so doing presumably experienced the pleasure of multiple erotic locations and identifications) now dreams that *he*—like Noah—had become the *object* of observation. By virtue of his changed spectatorial position, the Wolf-Man experiences the paranoid fear as the price to be paid for his earlier experience of spectatorial—and behavioral—satisfaction (Edelman 1994a).

By becoming an object in his son's eye, did Noah too experience what was, for the patriarch, a traumatic reassignment of spectatorial and identificatory position? Rather than being the disembodied eye, the panopticon who may see others but is himself not seen, one who stands above the "others," to be feared and respected, the one who "knows." "The disembodied voice," Kaja Silverman (1988, 164) writes, "can be seen as 'exemplary' for male subjectivity, attesting to an achieved invisibility, omniscience, and discursive power." (Silverman is thinking of cinema but the point is pertinent to the present context as well.)

Embodied, however, this panoptical spectatorial position collapses; now the father is a body (a "piece of ass"?), an object of another's gazes, feminized (castrated in Freud's term), an orifice waiting to be entered. Is it this the "castrating" spectatorial relocation that provokes the "paranoid fear" that the author(s) of Genesis 9:23 depicts as knowing "what his son had done unto him?" Like Freud's Wolf-Man, is Noah's paranoid fear—aggressively defended against by the curse—a cover-up of his guilt for the spectatorial satisfaction a father takes in watching his sons "come" of age? Such satisfaction is paternal not sexual, but does not the uncertain proximity of the two subject positions become discernible in the father's rage at even being looked at? Does not the longing underneath the rage surface in the fetish that is the foreskin?

THE HOLY FORESKIN

[H]ow do objects become attached to a desiring fantasy? (Teresa de Lauretis 1994, 300)

Sexuality is in itself a deviation, a departure from the real, from biology, from necessity, into the meandering detours of fantasy. (Elizabeth Grosz 1994, 281)

After all, men's bodies are like God's body. (Bjorn Krondorfer 1996, 4)

Martin Luther was amazed by the number of those who proclaimed possession of the holy foreskin. In the sixteenth century, abbeys from Antwerp to Bologna announced they held this piece of Christ's penis, a piece which had amazing healing powers. In Charroux, the relic was framed in silver; presumably it eased the discomfort of pregnancy and childbirth. A queen of Sicily, diagnosed with an incurable illness, made a pilgrimage to an Italian abbey were, she claimed, touching the foreskin of Christ healed her. At the same, numerous rumors circulated concerning nuns who abused the divine relic for sexual stimulation (Gollaher 2000).

Luther was a notorious anti-Semite (Morris 2001).[1] He was hardly the first. For centuries there had been rumors linking circumcision with the Christian hatred of Jews. In 1144, a gang of Jews was rumored to have kidnapped a young boy named William of Norwich. Presumably they shaved his head, tortured him, and cut his skin with thorns. Finally, according to chronicles of Thomas of Monmouth,

> they lifted him from the ground and fastened him upon the cross . . . . After all these many great tortures, they inflicted a frightful wound in his left side, reaching even to his innermost heart . . . . And since many streams of blood were running down from all parts of his body, then, to stop the blood and to wash and close the wounds, they poured boiling water over him. (Quoted in Gollaher 2000, 38)

On the chance that the meaning of this event might be lost on his readers, Thomas had one of the murderers proclaim: "Even as we condemned Christ to a shameful death, so let us also condemn the Christians, so that, uniting a Lord and his servant in a like punishment, we may retort upon themselves the pain of that reproach which they impute to us" (quoted in Gollaher 2000, 38).

More than a century later, in 1255, a Christian boy aged eight or nine named Little Hugh of Lincoln, was found murdered. He had been beaten, his nose broken, and his penis circumcised just before his death. Certain that only Jews could have committed the deed, Christians arrested ninety-one Jews; eighteen were executed. This was no isolated incident; during the period, there were a number of anti-Semitic practices and regulations. In 1243, for example, royal decrees reiterated the 1222 Council of Oxford, in which the construction of synagogues was forbidden. Sexual relations between Jews and Christians were also forbidden (Gollaher 2000).

Circumcision itself may be a sublimated sign of the "curse of Ham" in which the son is ritualistically marked as the father's property, but Christian fantasies of Christian boys mutilated at the hands of Jews suggest the pederasty embedded in the Eucharist. In the case of Anderl von Rinn (d. 1462), Christians imagined that Jews collected the child's blood in a bowl and used it to make Passover matzohs. The archetypal instance of this fantasy was recorded in late-fifteenth-century woodcuts illustrating the murder of Saint Simon of Trent. Just before Easter in Trento, Italy, in 1475, the body of a boy was discovered near the house of a Jew. All local Jews were arrested; eight were executed immediately, five later. Simon, meanwhile, was beatified and venerated as a martyr (until 1965, when the Roman Catholic Church withdrew the status). The prototypical image of Saint Simon's martyrdom, published in Hartmann Schedel's *Nuremberg Chronicle*, portrays Jews circumcising the two-year-old while they bleed him to death, presumably (according to Christian fantasy) saving his blood for use in their Passover ritual (Gollaher 2000).

Claudine Fabre-Vassas, a French ethnologist, described a Florentine engraving that depicts the martyrdom. As in many American lynchings, the focus was on the phallus: "The emphasis is placed on the treatment of [St. Simon's] genitals, which are being cut with a large knife. A gaping wound is opened at his throat, from which the blood is flowing into a receptacle . . . . Shearing scissors are ready to cut into his chest and needles prickling his skin contribute to bleeding him white" (quoted in Gollaher 2000, 39). This focus continued and even intensified, as in a seventeenth-century English fantasy recorded by Samuel Purchas:

> One cruell and (to speak the properest phrase) *Jewish crime* was usuall amongst them every yeere towards Easter . . . to steale a young boy, *circumcise him,* and after solemn judgement, making one of their own Nation a Pilate, to crucifie him out of their divellish malice to Christ and Christians. (Quoted Gollaher 2000, 40)

Does the "young boy" function as did the fragile white girl in late-nineteenth-century Southern white men's fantasies, the displaced site of desire disavowed as assault?

Such fantasies surfaced in legal proceedings, as in the court records of "the famous Trial of Jacob of Norwich, and Accomplices, for Stealing Away, and Circumcising, a Christian child" (quoted in Gollaher 2000, 40). In this case, court testimony reports that a five-year-old boy was abducted while playing in the street and taken to Jacob of Norwich's house. There his captors blindfolded him and cut off his foreskin. They then proceeded to play a game, the rules of which involved burying the severed foreskin in a basin filled with dry sand, then "blowing the Sand with their Mouths, till they found it again." The winner of the contest declared the boy a Jew (Gollaher 2000). At some point the boy was returned home and his kidnappers were brought to trial, in which his guardians told the court that "by some art or other" the circumcision had been reversed and the boy's foreskin restored (quoted in Gollaher 2000, 40).

In the eighteenth century, Johann Bodenschatz focused on the posthumous circumcision of infants. If a male baby died before the eighth day, he wrote, his foreskin would be removed, even in the coffin at the graveside, "so that he would not be buried with that emblem of shame or sin" (quoted in Gollaher 2000, 41). Other Christians spent time thinking about what the Jews did with the foreskin after circumcision. One popular legend held that the Jews buried the skin in sand so that a serpent might devour it, thereby linking circumcision to snake worship and, as well, to the ancient myth of rebirth and renewal symbolized by the snake shedding its skin (Gollaher 2000). The classic association of serpent with phallus and foreskin with the sign of castration also links circumcision with rape.

There were English churchmen and poets who explored circumcision's symbolic meanings in relation to the life of Christ. Gollaher points out that to the Puritans of the early seventeenth century, every aspect of the Old Testament, every subtlety of Mosaic law, constituted a foreshadowing of Christ. To illustrate, Gollaher quotes Richard Crashaw (1613–1659), an erudite poet who wrote "Our Lord in His Circumcision to His Father." Crashaw imagined the infant Jesus, on the eighth day, after his circumcision, addressing God-the-Father. In the poem, Christ twice sacrifices his blood to God-the-Father, first through circumcision, then through crucifixion. Circumcision not only presages crucifixion, it seems conflated with it, as the mohel's blade anticipates the Roman spear that pierced Jesus' side (Gollaher 2000). Is there a connection here between semen and blood, religious ritual and sexualized violence? Is the blood of Christ also his semen?

What if the Father did not curse the son, did not allow him to be crucified by other men, did not claim him as his property, or brand him with the mark of his "covenant"? What if Ham was given text time, enabled to report his experience of what happened inside his father's tent? Would we discover that the curse displaced the desire from the father to the son? What would be read if we heard the son's story? Would it read like the tale of a madman?

GOSPEL OF ST. SCHREBER

What I call Schreber's "own private Germany" consists of his attempts, using the available repertoire of culture values and valences, to interpret and to assign meaning to a maddening blockage in meaning that prevented him from assuming his place as a master of juridical hermeneutics and judgement. (Eric L. Santner 1996, 55)

Schreber's preoccupations reflect many of the preoccupations of his own day. They are filled with overt fear of emasculation and devirilization, expressed through the fear of becoming a Jew. (Sander L. Gilman 1993, 146)

[U]nregulated sexuality as a specific practice must be black. (Russ Castronovo 2001, 192)

What communion hath light with darkness? (II Corinthians 6:14)

Daniel Paul Schreber was born in Leipzig on July 25, 1842. The Schreber name is still known in Germany, primarily for small gardens named after Daniel Schreber's father. Moritz Schreber wrote extensively on public health, child rearing, and the benefits of fresh air and exercise; his work inspired gardeners in the late nineteenth century (for a gloss of Moritz Schreber's work, see Santner 1996, 89–90). He also devised instruments and regimens to prevent children from masturbating (and to help children break the habit), a calling not uncommon in America (see Pinar 2001, 400 ff.). This was during a time when prominent psychology textbooks reported that twenty-six percent of neurotic boys who masturbated ended up demented (see Laqueur 2003, 366).

For those who have studied the family, Moritz Schreber was the authoritarian German patriarch whose pedagogical practices and orthopedic devices (including those designed to prevent masturbation) may well have contributed to his son's psychotic breakdown. Schreber's older brother committed suicide in 1877; his three sisters all outlived him (Santner 1996). Eric Santner (1996, 47, emphasis in the original) suggests that Mortiz Schreber was "*more*" father than the ordinary father in that he embodied a "surplus" of "paternal power" and "authority." Such a larger-than-life father may well have contributed to his "transfiguration" in his son's "imagination" (1996, 47) from father to doctor to God. Schreber is the son who stayed inside Noah's tent.

Schreber began to study law in 1860, one year before the death of his father. After passing the state bar exam, he worked in several legal departments of the government, among them the civil administration of Alsace-Lorraine during the Franco–Prussian War and the federal commission charged with devising the new Civil Code for the Reich. He married Sabine Behr in 1878. Soon after, he ran as a candidate of the National Liberty Party (with the support of the Conservative Party), but lost to the socialist Bruno Geiser. This event was said to have triggered his first psychotic break, for which he was treated at the Psychiatric Hospital of Leipzig University under the care of its director, Paul Emil Flechsig. His primary symptom was severe hypochondria; this passed, as Schreber notes in his *Memoirs*, without any events "bordering" on the "supernatural" (1968 [1903], 62; quoted in Santner 1996, 3).

After his release from Flechsig's clinic, Schreber served as judge in several districts in Saxony; he was, evidently, healthy and contented, recording that after "recovering" from "my first illness" he enjoyed eight years with his wife, in general "quite happy," and "rich" in "outward honors," but "marred" by the "disappointment" of failing to have children (1968 [1903], 63; quoted in Santner 1996, 3). Schreber is referring here to the several miscarriages his wife suffered during this period (Santner 1996). Those "outward honors" led, in June 1893, to Schreber's appointment to the position of Presiding Judge of the Third Chamber of the Supreme Court of Appeals (Santner 1996). As time of his appointment approached, new symptoms appeared. He recalls:

> [O]ne morning while still in bed (whether still half asleep or already awake I cannot remember), I had a feeling which, thinking about it later when fully awaken, struck me as a highly peculiar. It was the idea that it really must be rather pleasant to be a woman succumbing to intercourse. (1968, 63; quoted in Santner 1996, 4)

The "man" to whom this "woman" would succumb, it turns out, was God-the-Father.

Schreber assumed his appointment as Presiding Judge in October 1893. Soon after, he began to experience additional symptoms, especially insomnia. During this first experience of sleeplessness he experienced "an extraordinary event." He recalls hearing a "recurrent crackling noise" coming from the bedroom, awakening him each time he was about to fall asleep. At first he assumed that the noises were caused by mice, but soon enough he was forced to a different conclusion, recognizing the sounds as "undoubted divine miracles" (1968 [1903], 64; quoted passages in Santner 1996, 4). This was the first event in what Schreber would come to characterize as an elaborate and divine conspiracy, a curse one might say.

By November 9, 1893—the day before the anniversary of his father's death—Schreber attempted to take his own life. This suicide attempt led to consultations with Flechsig, and Schreber was admitted, once again, to the university clinic where continued insomnia left him feeling exhausted and fragile. Several months into this second hospitalization Schreber deteriorated further. This turn for the worse had a sexual element, as Schreber notes: "Decisive for my mental collapse was one particular night; during that night, I had a quite unusual number of pollutions (perhaps half a dozen)" (1968, 68; quoted in Santner 1996, 5). It was at this time, Eric Santner (1996, 5) notes, that Schreber's paranoia positioned his psychiatrist at the "center" of a "vast" and "ultimately divine conspiracy." It turns out to be a conspiracy against his manhood.

At first taken to a private clinic, Schreber was moved to the Royal Public Asylum at Sonnenstein on June 29, 1894. There he remained under the care of its director, Guido Weber, until December 20, 1902. During his hospitalization Schreber was declared officially incompetent, a ruling rescinded after Schreber filed his own writ of appeal to the Supreme Court. Among the documents submitted to the court was the text of the *Memoirs*, which Schreber had composed by 1900 based on notes he had kept since 1897. After his release from Sonnenstein, Schreber published his *Memoirs* with Oswald Mutze, a Leipzig publishing house known for its list of occult and theosophical works (Santner 1996).

Upon his release, Schreber lived briefly with his mother and one of his sisters before returning to his wife in Dresden. Evidently unwilling to risk another miscarriage, in 1906 the Schrebers adopted a teenage daughter, Fridoline, who later reported that her stepfather was "more of a mother to me than my mother" (quoted in Santner 1996, 5). No longer a judge, Schreber did legal work for the family, including the administration of his mother's bequests upon her death in 1907. It was, evidently, a balanced, even leisurely, life for the released patient; he took long walks with his daughter, played chess and piano, and continued to read widely, including in Latin, Greek, French, English, and Italian. While his general well-being was punctuated by short fits of bellowing, he did not complain about his illness. It was his sister who reported that the voices that had plagued him for several years had become a constant, unintelligible noise. Sabine Schreber suffered a stroke in November 1907; within weeks Schreber was hospitalized for the third and last time, now at the new state asylum in the village of Dosen, outside of Leipzig. He remained there until his death on April 14—Good Friday—1911.

Among the symptoms recorded in Schreber's chart are outbursts of laughter and screaming, periods of depressive stupor, suicidal gestures, insomnia, and delusions of his own decomposition and rotting. Toward the beginning of the final section of his essay, "On the Mechanism of Paranoia," Freud introduces what Santner (1996, 52) characterizes as his own theory of "decadence." Freud claims, Santner suggests, that Schreber was compelled to experience directly the real "glue" (the word is Freud's) of nineteenth-century bourgeois society: "sublimated homoerotic desire" (1996, 52). Is this the desire the fear of which fueled the late-nineteenth-century crisis of European masculinity?

Even when sublimation succeeds, the sexual still threatens to surface, especially when "sublimation" is propelled by "projection," that displacement of disavowed subjective content onto what becomes, then, abjected "others." "The causes of his [Schreber's] delirium," Malcolm Bowie (1991, 109) argues, "may be traced back to an initial mispositioning of Subject and Other: the Other should be intrinsic to the signifying chain but has been moved to a position outside it." The parallel with the "threatening" black phallus, as white men have so often experienced it, seems obvious: the "other" originates as their subjectively experienced desire. What "remains"?

CONTRARY TO THE ORDER OF THE WORLD

The sexual member exhibited by the Christ Child, so far from asserting aggressive virility, concedes instead God's assumption of human weakness. (Leo Steinberg 1996 [1983], 47)

[I]ndividual sexual structuring is both an effect and a condition of the social construction of sexuality. (Teresa De Lauretis 1994, 303)

[R]acialization is constitutive of sexuality, and vice versa, in specific historical contexts. (Siobhan B. Somerville 2000, 165)

Schreber's concern over his decomposition is a recurring theme in the *Memoirs*; Eric Santner (1996, 6) characterizes it as "obsessive." (I will rely on Santner in this discussion of Schreber; Slavoj Zizek [1998, 172] characterizes Santner's reading of Schreber's *Memoirs* as "brilliant.") He points out that the metaphors Schreber uses to depict this literal and figurative decomposition resonate with a more general sense of decay, degeneration, and enervation, also evident among the fin-de-siècle intellectuals and artists Gerald Izenberg has studied.

Santner cites Max Nordau's treatise on "decadence" in the arts and culture, *Degeneration* (1892), a work that helped establish the term as a key metaphor for cultural decline, a metaphor later used by Nazi ideologues. A Hungarian-Jewish physician, Nordau wrote his French medical dissertation with Jean-Martin Charcot, the famous French psychiatrist. Later Nordau served as the vice-president of the first Zionist Congresses (1897–1903). Although he himself remained committed to a bourgeois faith in progress through knowledge, science, discipline, and strength of will, he was mindful of the loss of that faith among his contemporaries, especially among artists, writers, and intellectuals (Santner 1996; Gilman 1993; Boyarin 1997).

Nordau characterized the fin-de-siècle mood as "a compound of feverish restlessness and blunted discouragement" culminating in feelings of "imminent extinction," a sense of the "Dusk of Nations, which all the suns and all stars are gradually waning, and mankind with all its institutions and creations is perishing in the midst of a dying world" (quoted in Santner 1996, 6). Nordau argued that the constant vibrations undergone in railway travel were partly responsible for the shattering of men's nerves, emblematic of the approaching end of the established order. At much the same time, Charcot named the railway as a common site of nervous breakdowns (Mosse 1996). Schreber would feel *divine* vibrations.

Nordau was thinking, too, of the proliferation of newspapers and expansion of post services: "the humblest village inhabitant has today a wider geographical horizon, more numerous and complex intellectual interests, than the prime minister of a petty, or even a second-rate state a century ago" (quoted in Santner 1996, 7). George Mosse suggests that this lived experience of accelerated time and expanded space accompanying rapid technological progress became somatized and medicalized. Nordau felt certain that rapid and apparently uncontrolled bodily movement and restlessness were signs of shattered nerves, sure symptoms of degeneration (Mosse 1996).

For Nietzsche, Jonathan Crary (1999, 122) points out, decadence is synonymous with the "perceptual adaptability" demanded by "spectacular culture," a culture in which "one loses one's power of resistance against stimuli—and comes to be at the mercy of accidents: one coarsens and enlarges one's experience tremendously" (Nietzsche 1967, 27; quoted in Crary 1999, 122–123). For Nietzsche, Crary continues, decadence may extinguish the preconditions necessary for the emergence of new forms of life and invention. But Nietzsche did not view the problem as specific only to the late nineteenth century: "Decadence itself is nothing to be fought: it is absolutely necessary and belongs to every age and every people" (Nietzsche 1967, 25–26; quoted in Crary 1999, 124).

Nervousness and hysteria are key concepts here, conditions often diagnosed in men as neurasthenia, also the case in the United States at this same time (see Pinar 2001, 413). Laqueur (see 2003, 299) tells us that during the 1890s many Russian doctors thought that increased masturbation (not to mention prostitution and other sexual perversions) followed from rapid economic and political development. If onanism was the byproduct of bourgeois alienation and individualism, a "proper socialist, proletarian education . . . would wipe it out" (Laqueur 2003, 300). Or so Communist revolutionaries hoped.

The fact of the matter was, David Eng points out, that neurasthenia was associated with men marked by particular class deficiencies, that is, lower-class male laborers traumatized by industrialization. Trying to adapt the concept of male hysteric to a model of femininity, Eng reports that Charcot was puzzled by the appearance of hysterical symptoms in virile working-class men; it was assumed that hysteria should be found among the "effeminate" men of the upper classes—"homosexuals"—and "not among the strong and vigorous proletariat" (Eng 2001, 178). Although the most frequent cases of female hysteria occurred among the upper classes, the opposite was the case for men. "Male hysteria was," (Eng 2001, 178) reports, "most common among the working classes."

Perhaps class is too narrow an indicator of hysteria; perhaps "race"—while not employed then as a diagnostic category—is more suggestive in understanding the malady. Thomas DiPiero (2002, 2) suggests that white men are "nothing" if not expressions of "hysteria." In like terms, Mason Stokes

(2001, 163) declares: "Whiteness doesn't simply reassert itself in response to anxiety; rather, it is *anxiety* itself." In their earliest remarks concerning the illness, DiPiero (2002, 17) points out, Breuer and Freud suggested that hysteria amounted to a "failure" of speech: when the patient was unable to articulate to a "troubling event," that event "retained its affect" and "manifested itself in bodily symptoms."

In mainstream diagnoses of the malady of the day, it was the absence of self-restraint that led, presumably, to neurasthenia, even sterility, problems complicated and intensified by the use of "physical and moral poisons" (Mosse 1996, 82), such as alcohol, opium, and, of course sex, leading to debilitating sexually transmitted diseases (Mosse 1996; see, also, Bederman 1995). For Nordau, degeneration was accompanied by perpetual liminality or *interregnum*, a condition of "cultural fatigue" wherein the symbolic sphere—including social "forms, values, titles, and identities"—is no longer fully credible, no longer commands belief and, thereby, no longer "structures" the "life-worlds" of "individuals" and "communities" (Santner 1996, 6). Is "cultural" fatigue also a gendered crisis, as Gerald Izenberg's analysis suggests? (See chapter 5).

Eric Santner points out that the embodiment of the social ruptures that characterized modernity and the body's relative states of vitality or degeneration made the medicalization of cultural crisis possible. Medicalization promised mastery of what otherwise were complex and not easily managed modalities of social, political, and cultural malaise. Santner points to Nordau's emphasis on the dissolution of symbolic identities; he notes that it was as if scientific and medical knowledge could reinvigorate a weakened sense of social and cultural location, a sense of certainty as to one's position in a symbolic network. Although Santner will not emphasize this aspect, it seems clear to me that this symbolic network was, as well, a gendered location.

For Sander Gilman (see 1993, 153), it was an anti-Semitic and racialized location as well. (The idea that Jews were literally black was an "old" one [Wahrman 2004, 95]). Gilman suggests that, in his breakdown, Schreber sees himself as a Jew. In his unmanning he has become a Jew, but not any Jew: he has become the "Wandering Jew." He quotes Schreber's notation of the "foul taste and smell such impure souls cause in the body of the person through whose mouth they have entered" (quoted in Gilman 1993, 152). "The smell of the Jew and of the female," Gilman (1993, 152) asserts, "are both incorporated to provide the sexualized stench of the Jews' rhetoric." Not only the odor of women and Jews is discernible, apparently. Before I became gay, I recall a high-school (male) teaching colleague's claim (asserted with disgust) that he could "smell" a homosexual.

This attention to the senses points to at least one contemporary parallel to the situation one hundred years ago, namely a certain medicalization of cultural malaise and estrangement, an emphasis upon the biology of psychology and of gender (cf., the "gay gene" obsession). Students estranged from a school curriculum that offers few bridges among subjectivity, society, and academic knowledge are not only bored or inattentive; they are medically diagnosed and medicated. Eric Santner points out that the absence of certainty and strength of will and purpose associated with a secure positioning in one's cultural symbolic network is diagnosed by Nordau as part and parcel of a sort of a generalized attention deficit disorder, for, as Nordau suggested, "culture and command over the powers of nature are solely the result of attention" (1895 [1892], 56; quoted in Santner 1996, 8).

The atrophy of attention risks an overexposure of the mind to stimulation from *within*. "Distracted" by the flow of internal thought and feeling that is unrestrained by attention, brain activity can become capricious, apparently without aim or purpose. Through unrestricted free association—poststructuralists might praise the condition as the free play of the signifier—mental representations might surface into consciousness, and, in Nordau's political image, "are free to run riot there" (quoted in Santner 1996, 8). Such overstimulation can evidently produce an intense "feeling of voluptuousness," a state of bliss mixed with pain which Nordau links to "extraordinary decompositions in a nerve-cell" (quoted in Santner 1996, 8). Schreber himself becomes prey to unrestrained mental association and to feelings of "voluptuousness."

Men feeling voluptuous can be diagnosed as suffering from a form of "disease," but, as Sander Gilman (1993, 159) reminds, the disease from which Schreber suffered, Freud imagined, was the recently medicalized "disease" of homosexuality. "The Jewish God, the lower God, the brown God, raped Schreber," Gilman (1993, 159) points out. He quotes Schreber: "Fancy a person who was a [judge] allowing himself to be f . . . ked." Such vulgarity is, presumably, the language of what Schreber terms the "posterior" gods, Gilman (1993, 159) notes; it is, in the minds of many late-nineteenth-century Christians, "the crude language of Jewish sexuality." Once again, Gilman (see 1993, 159) quotes Schreber:

> The choice of the word "f . . . king" is not due to my liking for vulgar terms, but having had to listen to the words "f . . . k" and "f . . . king" thousands of times, I have used the terms for short in this little note to indicate the behavior of rays which was contrary to the Order of the World.

Such sexuality, Gilman reminds, was largely associated with Jews.

Like Gilman, Santner also links Schreber to anti-Semitism; he characterizes Gilman's scholarship as the "boldest attempt" at grasping Schreber's "preoccupation" with the "Jewish question" (Santner 1996, 108). Santner focuses specifically on specifically historical and biographical events, specifically Schreber's political affiliation with the National Liberal Party, among the most aggressive anti-Catholic forces during Bismarck's assault on papal authority in Germany. (Schreber had attempted, without success, to win a seat in the Reichstag on the Liberal Party ticket in the 1884 elections.) Those passages of the *Memoirs* concerning threats from Catholics and Jesuits are (paranoid) references, then, to actual historical events, as construed in the *Kulturkampf*. As Santner explains, after the stock market crash of 1873, anti-Catholicism shifted to anti-Semitism, and the same shift is evident in the

Memoirs. At certain points Schreber's political fantasies became preoccupied with the "Jewish question." The "mediating" or "transitional" term between these clusters of motifs, Santner (1996, 105) suggests, is "Slavism," a "signifier" enabling Schreber to shift the focus of his politico-religious preoccupations from Polish Catholics—a crucial target in the *Kulturkampf*—to Eastern European Jews.

Louis Sass (1994, 12) understands Schreber's malady as characterized by a "detachment" from "normal forms" of "emotion" and "desire," not a "loss" but, rather, an "exacerbation" of "self-conscious awareness." Like many such schizophrenic cases, he suggests, Schreber's life-world is more dominated by concerns less libidinal than cognitive or epistemological in nature. Sass (1994, 12) proposes that Schreber's madness is

> akin to Wittgenstein's notion of a disease of the intellect, born at the highest pitches of self-consciousness and alienation. Madness, in this view, is the end-point of the trajectory consciousness follows when it separates from the body and the passions, and from the social and practical world, and turns in upon itself; it is what might be called the mind's perverse self-apotheosis.

From my point of view, Sass' thesis represents a difference in emphasis, but not in substance, from Freud's and Santner's interpretations.

In a footnote, Sass (1994, 167, n. 5) acknowledges that he would not deny the importance of the work of Niederland and Schatzman, that postulates, for instance, that the "one" or "God" to whom Schreber refers is associated with his father, who continues to inhabit Schreber's experience as a "disguised epistemological introject." To appreciate the meaning of Schreber's case (and its various interpretations) for "race" requires that we entertain simultaneously, as intertwined, these emphases upon sexuality, power (and symbolic investiture), and tortured self-awareness.

At the end the nineteenth century in Europe, "jouissance" and the decomposition of cell tissue, Santner (1996, 8) reports, were associated with venereal disease, especially syphilis. This venereal peril was linked to the practice of prostitution, and in the United States thoroughly racialized, as many European Americans believed that most "Negroes" were infected. As in the United States, fears of sexually transmitted disease supported fantasies of bodies in various states of decay and decomposition, as well as horror of dementia and idiot progeny. But, as Santner (1996, 8) points out, syphilis was a "highly overdetermined disease formation," transmitting a complex of social anxieties and cultural meanings.

CHAPTER 5

Decadence, Disorientation, Degeneration

MODERNISM AND EUROPEAN MASCULINITY

What is the boundary line between the diehard assertion of rugged white male individualism and its simultaneous feminization and spectacularization? (Fred Pfeil 1995, 29)

Another word for the phallus was deus, which is etymologically related to our word for deity, and to the Italian word for leader, duce. (Paul Hoch 1979, 146)

[F]or the nineteenth century was an ever more visually centered age, when attitudes toward society and the nation were often expressed in aesthetic terms. (George L. Mosse 1985, 10)

[W]hite masculinity is less a thing, an entity, or even a position, than it is a response or a function. (Thomas DiPiero 2002, 231)

Between 1885 and 1920 in Europe, Gerald N. Izenberg (2000, 2) tell us, there was a "social and psychological crisis of masculinity." Izenberg explains how this crisis helped shape both the thematic concerns and the formal innovations of the early-modernist revolution in the arts. The centrality of gender to an understanding of modernism has been documented by feminist scholarly analyses of modernism's patriarchal "construction," or deformation, of women's identity, as well as modernist scholarship's erasure of women writers and artists (Izenberg 2000). In an era when men still largely dominated cultural production, shifting representations of the feminine were, in part, reactions to shifts, including "disturbances" in masculine identity (Izenberg 2000, 3). Although making "provocative" contributions to our "understanding" of modernism, Izenberg (2000, 3) judges, this scholarship also raises important questions of substance and method. After discussing these questions, Izenberg—in what seems to me a book of unusual precision and beauty—will suggest a different relation between masculine identity and modernist innovation. He does so by examining in some detail the interrelationship between the works and lives of three leading early modernists: Frank Wedekind, Thomas Mann, and Wassily Kandinsky.

Izenberg (2000) begins with an insightful commentary on modernism, masculinity, and method. Acknowledging that there is a large and growing literature debating the essence and scope of modernism, Izenberg (2000, 3) employs what he characterizes as a "somewhat standard, perhaps even conservative provisional definition." Modernism was, he explains, both formal or structural and substantive. It signaled a break with representation in the broadest sense, a rejection of the notion that art was the portrayal of the objective or "real." Modern artists lost confidence in official versions of reality, inscribed ideologically and in "realist" and "naturalist" art. Izenberg continues:

> The truth of personal incoherence behind the façade of autonomy and fixed social roles showed that psychic reality could not be adequately contained within the framework of conventional social identity, even where that framework took critical account of social conflict. Beneath the ideological optimism of modern capitalist materialism lay hidden a self ravaged by *suppressed longings* both instinctual and transcendental, longings contemporary society could neither account for nor satisfy. Modernists reacted to such perceptions by developing further earlier Symbolist ideas of stylistic autonomy and by using radical new forms not only to explore and express these ignored dimensions of subjectivity, but also, so they hoped, to answer its yearnings. (Izenberg 2000, 3, emphasis added)

Could these "instinctual" longings be decoded as sexual and racial? Did they structure the disorientation of positionality Lee Edelman theorizes?

Izenberg notes that his definition is incompatible with at least two other influential characterizations, one he deems older and no longer in fashion, the other recent and in vogue. The first, and older, definition is that modernism was a formal revolution in which the arts achieved, presumably, maturity by restricting their attention to their true subject, namely the elements of their own structure. These include: (a) the organization of line, color, and space on a flat surface; (b) the orchestration of pure sound severed from what Jonsson (2000) terms the expressivist paradigm; and (c) the deployment of language as a pure play of signifiers (Izenberg 2000).

The more recent and fashionable definition of modernism—with which Izenberg's definition is incompatible—is associated with the contemporary postmodernism. This interpretation associates early-modernist critiques of language, conventionality, and selfhood with postmodernist assertions regarding the infinite fragmentation of experience and of the subject, as well as the absence of any metaphysical grounds for truth. By this account, early modernists were concerned with recovering an authentic self that was imperiled or lost by uncovering its true instinctual and spiritual substrata, and, thereby, reuniting that self with the world and with truth. Izenberg (2000, 4) demonstrates that, at least for several canonical modernists, the self that was to be "recovered" and thereby "reconstructed" constituted a "truly masculine self," a subjective reconstruction achieved, "paradoxically," by men's "appropriation" of "ideal femininity." The strategy sounds not completely different from that devised by postbellum Southern white men in the United States.

The argument for a crisis of masculine identity in Europe at the end of the nineteenth century, Izenberg tells us, seems to have originated in the characterization of the (masculine) artistic and literary response to the emergence of the "New Woman" in Europe around the 1870s and 1880s, also cited as an important provocation for the U.S. "crisis of masculinity" (see Pinar 2001, chapter 6). First expressed as a category of literary analysis, the "crisis" thesis has been taken up by social and cultural historians to the extent that it has become commonplace of periodization in the history of European masculinity. As it is among U.S. historians (see Pinar 2001, 321), the thesis is not without its controversies.

There are contrasting characterizations of the crisis. To illustrate, Izenberg quotes Andrew McLaren, who argued that masculinity was radically restructured at the end of the nineteenth century:

New scientific norms of male and female sexuality were propounded in the late nineteenth century by sexologists and psychiatrists because social transformations appeared in the eyes of anxious observers to have undermined the explanatory powers of older notions of masculinity and femininity. (McLaren 1997, 2; quoted in Izenberg 2000, 5)

In contrast to McLaren, Izenberg notes, is George Mosse (1996), from whose history of European masculinity I will draw later. Mosse agrees that historical forces in the late nineteenth century threatened the notion of masculinity established in the eighteenth century, arguing that the new sexology[1] (among other developments) also succeeded in propping it up.

Despite these complexities, Izenberg believes it is possible to say just what idea of masculinity went into "crisis" at the end of the nineteenth century. It was, he tells us, a bourgeois refashioning of an older aristocratic ideal that, although significantly modified, was never completely dethroned. Until the mid-eighteenth century, the manhood ideal expressed the virtues of the medieval knight mixed with those of the early-modern aristocratic courtier; Izenberg (2000, 5) lists the following qualities: "courage, honor, military prowess, loyalty, and chivalry" as well as a certain "refinement of manners" and "liberality of spirit." He adds two qualities that, in his estimation, are insufficiently stressed.

The role of women in defining manhood, Izenberg asserts, was more central than even the notion "chivalry" implies. (In the American South, Mason Stokes [2001, 37] suggests, the "fear of female independence [was] disguised as southern male chivalry.") In Europe and in the American South, women's role consisted of being idealized and protected by men. Women personified the spirituality to which men aspired but could rarely achieve due to the worldly, that is, often violent, character of their manly obligations in the public sphere. In the eighteenth century, Izenberg continues, women were also regarded as prerequisites to a new standard of "sociability." As the presumably weaker sex unsuited for public life, women required men's protection. This position left many men with both respect and contempt for women; even so, with manliness imagined as the opposite of what was "womanly," success with women was required in order to measure up to true manhood (Izenberg 2000).

This "aristocratic" fantasy of femininity was interwoven with fantasies of ideal masculinity in another, more spiritual, way. The ideal knight was a Christian knight, and his accomplishment had to do not only with his devotion to his lord but to the Lord. As early-twentieth-century "muscular" Christianity in the United States would attempt to contradict (see Pinar 2001, chapter 5, section VIII), Christianity tempered, even feminized, the more brutal forms of masculinity by converting them to transcendent service to the master. Although, as Izenberg points out, men's spiritual yearnings were culturally cut off from the warrior and restricted to both women generally and to that particular subset of men whose special obligation was the care of men's souls, the religious foundation of aristocratic manhood remained undisturbed.

By the middle of the eighteenth century, however, middle-class writers were translating aristocratic codes of masculinity into bourgeois virtues. Aristocratic codes were, Izenberg tells us, not so much challenged as much as simply supplemented, transmuted, or appropriated. Productivity and economic efficiency replaced aristocratic idleness; bourgeois domesticity and morality covered over aristocratic libertinism. It was at this time that masculinity came to be associated with an ideal body type derived, as we will see in this chapter, from the classical Greek ideal: strong and muscular but, also, sculpted, balanced, and serene. This ideal masculine body was, presumably, expressive of the ideal bourgeois masculine character; it was both the achievement and the badge of self-discipline, moderation, and civilization. This new bourgeois concept of manhood was not, Izenberg asserts, opposed to the warrior ethos of aristocratic manliness but, instead, a modern extension of it: what modernity offered was a less destructive, more productive site for manly initiative, courage, and daring. Rather than on the battlefield, the new bourgeois "warrior" fought in the bloodless tournaments of the marketplace and political public sphere (Izenberg 2000).

Bourgeois manliness was balanced between aggressiveness and discipline; the "authentically manly man," writes Peter Gay (1993, 103; quoted in Izenberg 2000, 6), was simultaneously "self-assertive" and "self-controlled." It was, presumably, self-control which spelled the difference between the bourgeois and both the new urban lower-class man, whose instinctualism, even animality, meant the threat of indulgence or violence, and the dissipated aristocrat (Izenberg 2000). Self control was key to bourgeois manhood in the United States as well (see, for instance, Bederman 1995).

Recall that, in the United States, class distinction was also racialized. European Americans imagined African Americans as the European bourgeoisie imagined the lower classes, as lacking "civilization." Born a slave, Ida B. Wells's success in persuading English audiences (during her lecture tours in 1892 and 1894) that it was the lynch mobs, not the alleged black male rapists they castrated, who lacked manly self-control was, in part, a consequence of her performance of that "dignity" and "self-restraint" her listeners associated with "civilization" (Bederman 1995; Ware 1992; Pinar 2001). It is a self-conferred dignity in the service of pedagogical activism that might inspire contemporary educators, themselves victims of displaced and deferred misogyny and racism (Pinar 2004a).

In late-nineteenth-century Europe, the construction of the middle-class businessman and professional as combative functioned to appropriate the aristocratic warrior ethic for the bourgeoisie. As Izenberg observes, physical prowess remained an important element in the ideal of masculinity in the so-called bourgeois era. The duel was still sanctioned in Europe into the early twentieth century as the "manly" means to defend one's honor (Izenberg 2000). As we have seen in the case of the American South, honor as well as manhood were statuses earned and protected in the company of other men, an "ideal" to be attained through struggle (DiPiero 2002; Gilmore 1990). It reflected and required a conception of (not only) women as "opposite" and "separate" (see Pinar 2001, section II).

The separate-spheres ideology influenced all major institutions in the West, among them the church and the school (see Haynes 1998; Tyack and Hansot 1990). It was as well a fact of economic life, but the rationale for this separation—that women were fragile creatures unfit for life in the market-place—was devised, in part, to keep women out of the affairs of business in order to preserve that domain of male autonomy. Aesthetic or high culture—that trace of the aristocratic ideal of manliness—was, like religion, relegated to the woman's sphere. Both realms were conceived as elevated and spiritual spheres that "ennobled" men trapped in "lower spheres" of economic activity. Wives' cultural and religious activities elevated their husband's social standing and their masculinity even as these remained quintessentially feminine domains. Because they—both aesthetic culture and the women who cultivated it—were confined to the domestic sphere, men could remain manly while being associated with higher pursuits. Women presumably enjoyed that spiritual inwardness which rendered them the profoundly human creatures men said they themselves aspired to be but which their worldly striving—for the sake of the family, of course—made all but impossible to achieve. One consequence was the bourgeois and, later, racialized, ambivalence about culture and education (Izenberg 2000; Hofstadter 1962; Dance 2002).

It was this "bourgeois" ideal of masculinity that, many historians have argued, fell into crisis toward the end of the nineteenth century. The evidence for this crisis is derived from disparate sources: Izenberg (2000) lists: (a) late-nineteenth-century economic developments that presumably weakened middle-class economic roles; (b) political and social challenges by women to previously exclusive male bastions of power, such as politics; (c) widespread concern over the presumed decline of virility due, it was argued, to the soft-ness of modern urban commercial and consumer society; and (d) the challenges to masculinity posted by the appearance of modern sexology, with its interests in bisexuality. Perhaps in this last category Izenberg would include the appearance of the figure of the "homosexual," marked in England by the trial of Oscar Wilde (Dollimore 1991; Pine 1995).[2]

None of these sources, Izenberg argues, offers direct evidence of a subjective sense of crisis as experienced by middle-class European men. What evidence there is of internal feelings comes primarily from literature and the arts, Izenberg notes, but even this evidence, he continues, seems compromised, due to a certain circularity in argument. Evidence for the crisis-of-masculinity thesis, he laments, is sometimes inferred from the work of artists and then explained, in circular fashion, by locating these feelings in economic and social developments. Of course, given the complex reciprocity between inner feelings and "outer" reality, a certain circularity in theorization of the "crisis" seems inevitable.

One feature of the social and historical reality that, no doubt, contributed to inner feelings of crisis was the rise of large-scale social and political movements on the Right and Left in the last quarter of the nineteenth century in Europe. The age of mass politics that followed the extension of the franchise in the fifteen years after 1870 ended, Izenberg notes, the upper-class monopoly of politics. He notes Carl Schorske's argument, in his study of fin-de-siècle Vienna, that this development prompted especially middle-class writers and intellectuals to abandon hope of exercising power in the public sphere and turn anxiously instead to an exploration of the psychic interior and its tensions and complexities. (Robert Musil's oeuvre for instance, *Young Torless* [1955 (1906)] could serve as one illustration.) Although Schorske's work does not deal explicitly with a crisis of masculinity, Izenberg observes, his analysis has obvious implications for the male identities of his figures, implications that have more recently been explored by Le Rider (1993) and Toews (1997). Even so, Izenberg points out, indisputable historical facts—such as the rise of mass political movements on the Left and Right and their assaults upon bourgeois liberalism—tell us nothing concrete about the subjective experience of middle-class men.

Izenberg cites other historical developments that might have contributed to an European crisis of masculinity, among them the poor performance of British soldiers against the Boer farmer militias, an event, he notes, that provoked anxiety over the presumably sapped virility of men subject to modern civilization. Among the responses in Britain was the Boy Scout movement (and in the United States the proliferation of fraternal orders: see Pinar 2001, chapter 6). Fear of depopulation in France as birth rates declined in absolute terms or fell below those of national adversaries provoked anxiety over a presumed decline of masculinity. But these expressions of anxiety over modernity's impact on manliness were interpretations and speculations made by observers worried about national strength, not testimonies from those who were in fact undergoing anxiety over their imperiled masculinity (Izenberg 2000). The crisis of masculinity these observers recorded was, Izenberg (2000, 9) notes, "always somebody else's."

More persuasive but still, in Izenberg's (2000, 9) judgement, only "indirect evidence for a widespread sense of internal anxiety about masculinity" is found in the systematic efforts to categorize masculinity, and in so doing split off and isolate the "unmanly" behavior of some men and the "unladylike" conduct of some women through "scientific" schemes that pathologize "abnormality." The popularity of "sexology," not to mention the appearance of psychoanalysis in Vienna, suggests that there was considerable anxiety over men's performance of masculinity (Izenberg 2000). Evidence from men's subjective experience would seem to be key, Izenberg (2000, 12) argues, in depicting a "crisis of identity" that was a "crisis of subjective consciousness." For this reason, Izenberg continues, works like Otto Weininger's *Sex and Character*, while not autobiographical, hold significance for understanding the crisis. Much of the subjective as well as "objective" evidence of a crisis of masculinity in fin-de-siècle Europe derives from or is about intellectuals, writers, and artists, whose occupations require some degree of self-consciousness and self-observation. Gendered notions of the "demonic woman" or "femme fatale," Izenberg points out, have been drawn from portrayals in literature, painting, and opera in the works of Wilde, Beardsley, Strindberg, Klimt, Wedekind, and Strauss, among others (Izenberg 2000).

If businessmen felt uneasy over the manliness of their occupational pursuits, artists felt even more so. Men who were artists were unable to defend the virility of their work by claiming material usefulness, wealth production, or the combative fortitude the marketplace demanded. An artist's gendered struggle for acknowledgement as "man" took place along two fronts. On the first, he fought alongside his fellow bourgeois, for recognition of their mutual claim to manliness against the old warrior ideal. On the second, he fought against his fellow bourgeois, forced to defend the idea that artistic creativity was manly and productive labor in a materialist culture (Izenberg 2000).

In the cases of the three artists he studies—Frank Wedekind, Thomas Mann, and Wassily Kandinsky—Izenberg focuses on both aesthetic production and lived experience. Such an "intertextual" examination "substantiates" (Izenberg 2000, 16) the claim that these artists' crises of masculinity involved their relation to femininity. But in none of these men's lives did femininity function simply or exclusively as a nostalgic notion of pre-individuated "being," even when it was associated with the maternal. Very much the contrary was the case, Izenberg tells us, especially for Wedekind and Kandinsky. Mann's case he judges as somewhat more complex. For these modernists in crisis looked at "the feminine" as representing "both autonomous creative power and connection with the whole of being, the union of the best of modernity with the best of premodernity" (Izenberg 2000, 17).

In contrast to prevailing stereotypes of femininity, Izenberg argues, these artists seemed to associate the ideal feminine with Nietzsche's *Ubermensch*, the human being of the future. Freed from dependence upon obsessions with the transcendental, the "ideal feminine" was content to dwell in her body, in this world, for which she legislated her own laws, fashioned her own forms, as Zarathustra had taught. These artists were expressing "not a nostalgic yearning for a regressive return to undifferentiated fusion, a search for ultimately security that meant dissolution of the self, but a quest for a restoration of their own creativity through an appropriation of the feminine" (Izenberg 2000, 17).

To use patriarchal biblical imagery, is this not an effort to restore Adam's rib? Is it in this sense that "woman" "*n'existe pas*"? For Wedekind, Mann, and Kandinsky, Izenberg argues, this appropriation of the notion of the feminine was also rather un-Nietzschean. For these men would seem to have internalized something of the "separate ideology" idealization of the feminine, as its appropriation, Izenberg (2000, 17) tells us, "would mean not just human self-sufficiency, but godlike wholeness." If the rib were restored, does Adam become his Father?

The subjective sense of imperiled masculinity that prompted the artists' quest for a restoration of masculine power derives, Izenberg notes, from a mix of individual and social factors, not the least important of which was an undermining doubt each felt about the value of art itself. In the late-nineteenth-century European crisis of masculinity, many men seized upon an idealized femininity as a tactic toward the "restoration" (as some nostalgically imagined it) or rejuvenation of masculinity, a tactic, Izenberg argues, which only exacerbated their sense of masculine vulnerability and led to a profound ambivalence toward femininity.

Why? In order to restore their masculine and creative potency, these artists felt they must incorporate femininity into their beleaguered selves. But in doing so they became unnerved. What if they were overwhelmed by this powerful femininity? What if they disappeared in the incorporative process, losing not only their autonomy but their very (masculine) selves? This ambivalence was intensified by the contradictory but powerful fact that these artists also retained the dominant cultural image of femininity as weak, passive, and dependent. Here leaning psychoanalytically himself, Izenberg (2000, 18) suggests that this was "partly a defense against their own idealization of the feminine." But, he continues, the defense failed as it associated them with traditional stigmas of feminine weakness and passivity, the very vexed condition they were working to surpass. The aesthetic enterprises of Wedekind, Mann, and Kandinsky, Izenberg argues, can thereby be comprehended as a series of strategies for restoring and rejuvenating masculinity by incorporating the feminine while negotiating the ambivalence that a project so conceived necessarily produced. It is clear to Izenberg that none of the three artists succeeded, either logically or existentially, in reconciling the inherent contradictions of this gendered project. Aesthetic success, he notes, depends neither upon logical nor psychological success, observing simply but dramatically: "Their efforts produced an aesthetic revolution" (Izenberg 2000, 18).

We turn next to a tale of theological and epistemological revolution. In contrast to Izenberg's narrative, this story is not one of masculine rejuvenation through the appropriation of the feminine, but one of the rejection of patriarchy by becoming feminine. Unlike Noah, Daniel Paul Schreber abandoned the patriarchal subject position by succumbing to the desire of the Father. (In Christianity, the distinction between Father and Son blurs on the totem that is the cross.) Like the subjects of Izenberg's study, Schreber, too, was living through fin-de-siècle decline and decadence, but, as Eric Santner (1996, 52), points out, Schreber's "analysis" of decadence differs from those of other theorists of cultural decline in its "distance" from the "symptoms" of degeneration. Schreber's analysis is, Santner (1996, 52) notes, "inseparable" from his "perverse" living out of these "symptoms." Schreber was, in my view, a nineteenth-century Noah, a patriarch who performed, not repudiated, his vulnerability, his lack, his effeminization in face of his Father's desire. This time there would be no sons cursed to servitude except, of course, for the son come of age: Schreber himself.

Soul Voluptuousness

"Thou shall love the Lord thy God with all thine heart, and with all thy soul, and with all thy might." Why does this love have to be commanded? And why does it so often fail? (Regina M. Schwartz 1997, 113)

I have a complicated history of bodily relatedness with Jesus which informs my gaze upon the man Jesus from the foot of the cross where his naked, dead body hangs. (Robin Hawley Gorsline 1996, 125)

What kind of man am I, and what kind of a man do I want to be if I want to be a man at all? (Daniel Boyarin 1997, ix)

During his hospitalization, Schreber became convinced that God wished him to be transformed into a woman. According to the theology Schreber formulated as a result of his experience, the soul is located in the nerves of the body and after death returns to God, who himself is comprised of nerves. Upon returning to God, the soul-nerves undergo purification and attain a state of blessedness in which there is uninterrupted enjoyment. Sounding here like a pre-AIDS gay liberationist, Schreber believed that the soul's happiness was the experience of continual pleasure, a state he identifies with the "feminine." When describing his own transgendering into a woman, Schreber employs the phrase "soul voluptuousness." Recall that Schreber believed there is an upper and lower God; it is the rays of the lower God, Ahriman, that can "unman" human men (Schreber 2000 [1903]). This implies not only feminization but, more precisely, the "self-shattering ecstasy" that Leo Bersani (1995) associates with anal intercourse and that Elizabeth Grosz (see 1995, 109) associates with lesbian sexuality. Schreber felt certain that the end of the world order was imminent and that he alone would survive the cataclysm. His metamorphosis was not limited to gender; he also regarded himself as the eternal Jew, who, once unmanned, would become the means by which the species will be reproduced. As Sander Gilman (1993) has pointed out, late-nineteenth-century Jews were often feminized by Christians and other non-Jews. Schreber's becoming woman—from the removal of his mustache to the "change of my whole stature"—"emanated from the lower God (Ahriman), the god of the Jews" (quoted in Gilman 1993, 155).

Eilberg-Schwartz reminds us that initially Schreber imagined that his psychiatrist, Dr. Flechsig, was responsible for his unmanning, but soon enough it became clear to him that it was no less than God-the-Father who was "behind" it. For the late-nineteenth-century German judge, being the object of his Father's eye meant being unmanned, for him not only a psychological state. In fact, Schreber reports changes in his body, including increased "soul voluptuousness" and indications that his male organ was retracting: "The 'soul voluptuousness' has become so strong that I myself received the impression of a female body, first in my arms and hands, later in my legs, bosom, buttocks and other parts of my body" (1968 [1903], 148; quoted in Eilberg-Schwartz 1994, 35). Initially horrified by his unmanning, Schreber soon enough regards his new "positioning" as his right and duty.

It was C.G. Jung, Sander Gilman (see 1993, 141; see also Santner 1996, 22) points out, who suggested, in the spring of 1910, that Freud read the autobiography of Daniel Paul Schreber, which had appeared in a limited edition in 1903. Earlier, Jung had used this material in *The Psychology of Dementia Praecox* (1907). Freud felt certain that Schreber's memoirs illustrated his theory of paranoia in that they revealed the inability to work through what Freud conceived, at this stage, as a contradiction: that I (a man) love (a man). Like many nineteenth-century Southern white men, Schreber, Freud theorized, suppressed the wish, which was then reversed, surfacing as an apparently external event: "I love this man" becomes "This man wants to harm me." (In the case of nineteenth-century Southern white men, the wish surfaced as "this man wants to rape my daughter [wife, mother, sister]"), the latter an abstract and transgendered site of regressed and desiring identification.

Schreber's paranoia was provoked, Freud claimed, by the surfacing of homosexual desire while his wife was away on vacation, a micro-version of what mid-twentieth-century U.S. prison researchers imagined as "situational homosexuality" (see Pinar 2001, chapter 16). As evidence, Freud cites Schreber's report that the onset of the illness was accompanied by twelve emissions in one night. The presence of Schreber's wife, Freud speculated, served as a substitute for the men he desired; her absence precipitated an eruption of homosexual desire. Eilberg-Schwartz (whose summary I am following here) is especially interested in Freud's argument that for Schreber God is a substitute first for Flechsig, himself a substitute for Schreber's father. When Schreber asserts that God understands nothing about living men, that he knows only how to deal with corpses, he is, in effect, speaking about his own father (Eilberg-Schwartz 1994).

Schreber does not only attack his father-God, he desires him, but this desire is, apparently, reversed, as Schreber imagines that it is God who desires him. (Did Noah "reverse" his desire by cursing the son he raped?) As Eilberg-Schwartz reminds, Freud suggests that God permits Schreber to disguise and thereby experience his own desire for his father. Given the heterosexual imperative, God-the-Father demands that I (his son and lover) be transformed into a woman. Schreber's fear of and desire for castration by his father is experienced as his involuntary transformation into a woman, a desire the genesis of which is relocated to God (Freud 1911; Eilberg-Schwartz

1994). Speaking of Schreber but as well, it seems to me, of many men who worship (a male) God, Eilberg-Schwartz (1994, 36) concludes: "God, then, is a concept through which repressed homosexual feelings are finally acknowledged and accepted." The sacred and the sexual appear here as two sides of the same male coin.

Eilberg-Schwartz acknowledges that Freud's reading has been criticized from several points of view, and that, specifically, his theory that behind paranoia lies repressed homosexuality is no longer taken seriously. Some have suggested that Schreber's breakdown was a predictable consequence of childhood abuse, although this interpretation need not necessarily contradict Freud's (Schatzman 1973; Eilberg-Schwartz 1994). Others decline to dismiss Schreber's memories as the ravings of a psychotic man, discerning in them a sophisticated critique of gender relations (e.g., Schreber's interest in becoming a woman), a theological critique of a God unaccountable to any ethical obligations outside those he himself recognizes, and an epistemological critique of the symbolic system of late-nineteenth-century Europe (Geller 1993, Eilberg-Schwartz 1994; Sass 1994, Santner 1996). Freud's theorization of Schreber's memoirs interests Eilberg-Schwartz because it discloses that Freud was well aware of the son's homoerotic relationship with his father and its expression in the concept of God, an insight with, Eilberg-Schwartz (1994, 37) suggests, "more general applicability."

Schreber's fantasies of being unmanned and sexually desired by God, Eilberg-Schwartz continues, took him to the center of what Eilberg-Schwartz terms the dilemma of monotheism. Schreber, he writes (1994, 137), was able to "think the unthinkable" and in so doing express what traditional theology has refused to think. As Philip Greven (1977) made clear in his study of Protestants in America, Eilberg-Schwartz asserts in the context of Judaism: "When a man confronts a male God, he is put into the female position so as to be intimate with God" (1994, 137). It is this "gender trouble" that results, he argues, in the prohibition on graven images.

The masculinity of ancient Israelite men was evidently unchallenged when God-the-Father turned his back, hid his face, or kept himself covered in a cloud or in the heavens. But when these men had to face their (male) God, men's masculinity was challenged. (Why not God's?) In ancient Judaism, Eilberg-Schwartz suggests, this divine destabilization of masculinity proceeds in ways parallel to Schreber's unmanning. It occurs on occasion, Eilberg-Schwartz (1994, 138) points out, through "violence" that "threatens castration," even "death," and on other occasions through "more subtle forms" of "gender reversal." (The intimacy of Israelite men with their male God involved "feminization, loss of manhood, and perhaps even death," reminds Eilberg-Schwartz 1994, 151.) There was, then, Eilberg-Schwartz (1994, 151) asserts, a "danger of intimacy with a male God, a threat that could be diminished only by a partial unmanning of the Israelite man." Stripped of its meaning as religious ritual, circumcision constitutes the mark of sublimated sexual ownership of the son by the father. Is the "sacred" what remains after homosexual desire is subtracted?

The gendered dynamics of men's relationship to a male God suggest so. Given the Leviticus laws and ancient Judaic binaries, to be a devout man meant imagining oneself as a "woman," at least when the analogue of the divine/human relationship is marriage, as it was in ancient Israel and continued to be in late-antique Judaism and even through eighteenth-century Protestant Christianity. This positioning of men as women is, Eilberg-Schwartz points out, undeveloped in Scripture, but was articulated by late-antique rabbinic interpreters of Judaism. He asserts that the rabbis understood that in the relationship with God, men must assume the position of wives, a position Daniel Paul Schreber would assume. On occasion, Eilberg-Schwartz (1994, 163) reports, the sages read Scripture "as if they imagined themselves as women," often looking to women for "guidance." Such self-positioning was not "social only," he continues, but "eroticized, as seeing God was a decidedly erotic experience" (1994, 163). Amen.

MODERN MASCULINITY IS A STEREOTYPE

Rarest among Infancy scenes that refer genitally to the manhood of Christ are images of the Child with the penis erect. They survive in sufficient number to testify that sixteenth-century painters and patrons thought the motif not inappropriate. (Leo Steinberg 1996 [1983], 76)

The engendering of a young male as culturally masculine holds him in the thrall of a father figure. He is enjoined to identity with that figure while at the same time forbidden to emulate him absolutely. (Thomas DiPiero 2002, 34)

Masculinity is a symbolic construct at odds with itself. (Howard Eilberg-Schwartz 1994, 16)

Modern masculinity is a "stereotype," George Mosse (1996, 5) tells us, a standardized ideal, the "unchanging representation of another," as Webster's Dictionary defines the word. As a stereotype, modern masculinity obliterates individuality, replacing it with the cult of individualism in which individuality is replaced by conformity, a rationalization for social and economic exploitation, but that story I save for another day. As stereotype, modern masculinity may be peculiarly susceptible to stereotyping others, and not just along gender lines, but along racial and class ones as well. George Mosse (1996, 5) asserts that the stereotype of masculinity is a composite "based upon the nature of man's body, but as psychoanalytic theory would suggest, the body is imagined first and experienced later." In other words, the stereotype follows from a fantasy of an "ideal" body circulating in one's own libidinal economy, often a visual fantasy (see Chow 2002, 53). The disjuncture between the fantasy and reality is what, for DiPiero (2002, 185), constitutes the "crux" of "white masculinity's ideological grip." Both at the level of culture and the individual psyche, he points out, "no one" coincides with the ideal of white masculinity, making "failure inevitable" and, thereby, valorizing all the more intensely the fantasy.

As ideals producing conformity, patriotism and masculinity are associated. Co-extensive with the ascendancy of political imagery—for instance, the national flag—as patriotic symbols, Mosse tells us, was the ascendancy of the symbolic meaning of the human body itself. In this sense, Mosse (1985, 41) asserts, "racism was a heightened form of nationalism." During the second half of the eighteenth century, monotheistic Europe became "more visually oriented," Mosse (1996, 5) concludes, evident not only in national symbols such as flags, but also in "sciences" such as physiognomy and anthropology, with their classifications of men according to standards of classical beauty. The masculine stereotype became integral to "an ever more visually centered age" (Mosse 1996, 5). As the human body assumed symbolic form, its construction and its beauty took on increasing cultural and political importance. "Preoccupation with the human body," Mosse (1985, 178) points out, "was typical for fascism as a visually centered ideology, an attitude toward life based on stereotypes."[3]

The stereotype of the "masculine" was strengthened, Mosse argues, by the presence of a binary, that is, negative stereotypes of men who not only failed to coincide with the ideal but who, in body and soul, defined the exact opposite of true masculinity. Groups marginalized by society, including Jews and blacks, functioned as these stereotypes. Mosse (1996, 6) describes racism as "based upon stereotypes and stereotyping." The presumably misshapen bodies of Jews (who tended to be feminized) and blacks (who tended to be hypermasculinized) were, presumably, the signs of their racial degeneration. To illustrate, Mosse cites Friedrich Ehrenberg, a leading German Protestant clergyman, who asserted that the ideal of the true man must be kept clearly in view as a kind of compass, guiding aspiring young men away from those who were "immoral, weak and servile" (Mosse 1996, 6). Here we see the curse of Ham in the gendering of European culture.

Mosse (1996, 6) emphasizes the public character of a stereotype. The public nature of gender stereotypes, he writes, "made the invisible both visible and public, and it was in this manner that stereotypes gained their social and political importance" (1996, 6–7). To illustrate his point, Mosse quotes from a late-nineteenth-century English phrenological magazine: "man may be considered in the light of a placard, hung up on the wall to be read," and "our virtues, vices, excellences, culture or barbarism, can be seen by those who have eyes sufficiently educated to read and understand their external manifestations" (quoted in Mosse 1996, 7), a correspondence Frederick Douglass exploits (see chapter 4 "Contrary to the Order of the World"). Even when masculinity was associated not with the body but with abstract morality, it was, Mosse asserts, based still upon certain standards of appearance, behavior, and comportment.

At the beginning of the nineteenth century most European women lost, Mosse tells us, those small gains made during the eighteenth-century Enlightenment. In Europe as in North America, a "separate spheres" ideology reigned, and women were confined to a domestic sphere clearly distinct from the public one assumed to belong to men. As we have seen in the case of conservative Christianity (see Haynes 1998; Pinar 2001, chapter 5), this division did not mean that women were inferior to men, only that they had different functions, ordained, presumably, by God, biology, or evolution. Men and women were imagined to be "opposites," complements to each other (Mosse 1996). (St)Eve lives.

This difference was key in the construction of modern masculinity, which, as Mosse explains, defined itself against countertypes, feminized (Jewish) or hypermasculinized (black) men, against, in object-relations terms, their pre-oedipal identification with the maternal body. This latter defense was symbolized as the "difference" between "opposite" sexes. In object-relations terms, when the maternal body gets abstracted, it becomes the "other" and is imagined as opposite the man. Men who were not so obviously "opposite" to women became known as effeminate, a term that came into general usage during the eighteenth century. It denoted an unmanly softness and femininity (Mosse 1996). It was often associated with sodomy.

To illustrate what he takes to be an increasing significance of the visual in the modern period, Mosse recounts Johann Kaspar Lavater's theory of human physiognomy. Lavater introduced, Mosse tells us, a new way of seeing men and women. Whereas traditionally clothing held the "measure of a man," now it was their physical profile: the shape of the nose, the color of the eyes, and bodily structure. These characteristics conveyed, presumably, a man's true character. This new "science," which Lavater formulated in his *Essays on Physiognomy* (1781), was based on "the ability to recognize the hidden character of a human being through his outward appearance" (quoted in Mosse 1996, 25). J.J. Winckelmann provided Lavater with the standard by which appearance could be gauged: the ancient Greeks exemplified ideal human beauty, and such beauty, presumably, embodied and expressed true morality, a morality, of course, that included erotic relationship between older and younger (distinctly underaged in contemporary legal terms) men and boys. Echoing Winckelmann, Lavater declared that the Greeks were morally superior, not to mention more beautiful, than the present generation (Mosse 1996).

This was, of course, no free-floating aesthetic judgement; nor was it only a thinly disguised longing for a more openly homoerotic historical moment and social structure. In this conflation of morality with beauty was, Mosse (1996, 25) argues, a "certain pragmatism, an emphasis on the material, combined with concern for the proper morality." As Lavater formulated the equation: the more virtuous a human being, the greater his/her beauty; the less virtuous, the uglier his/her appearance. Physiognomy was important in the production of modern masculinity—and in the production of race, I might add—because it linked the body and soul, morality and bodily structure (Mosse 1996).

From his youth onward, archeologist and art historian Johann Joachim Winckelmann (1717–1768) was "obsessed," Mosse (1996, 29) tells us, with the beauty of Greek sculpture, an art form that had been neglected for centuries in Europe. In his most influential works, *Reflections on the Painting* and *Sculpture of the Greeks* (1755) and the well-illustrated *History of Ancient Art* (1764), Winckelmann sought to legitimate an ideal of beauty based on descriptions of ancient Greek sculpture, insisting upon the primacy of sculpture over the other arts. On this point Winckelmann's contemporary, the philosopher Johann Gottfried von Herder, echoed Winckelmann that only sculpture could represent multidimensional truth. Painting, in contrast, amounted to a "narration suffused by magic" (Mosse 1996, 29). Only through sculptural representation of the details of the male body could the viewer appreciate "ideal" male beauty. This "ideal" male beauty became, in the modern period, George Mosse (1996) argues, a male stereotype.

The sculpture upon which Winckelmann focused as representational of ideal beauty was primarily that of young male athletes whose muscled bodies exemplified, Winckelmann argued, power, virility, harmony, proportion, and self-control (Mosse 1996). The "noble simplicity and quiet grandeur" of these sculptures dazzled Winckelmann, whose use of such memorable phrases added to the appeal of the argument (quoted phrase in Mosse 1996, 29). The youthful male bodies that he described were, Mosse tells us (1996, 29), "always lithe, without any surplus fat, and no feature of the body or face disturbed their noble proportions." Such an "ideal" body communicated, presumably, both strength and restraint, the balance that Lavater had also praised, illustrated by a sculpture of Apollo of Belvedere, regarded by many as the most beautiful of the young Greek gods. As Goethe was to write in 1771, "Apollo of Belvedere, why do you show yourself to us in all your nakedness, making us ashamed of our own" (quoted in Mosse 1996, 32). Hear the echo of Noah here?

Why did Winckelmann formulate his ideal of beauty as balance, proportion, and moderation, three qualities that for him reflected a great and tranquil soul? The answer, George Mosse (1996, 33) speculates, is that Winckelmann was under the influence of that "cosmopolitan harmony" advocated by the Enlightenment. Perhaps personal factors also played a role; perhaps Greek sculpture provided order in his own chaotic life. "[C]ertainly," Mosse (1996, 33) continues, "his homosexuality may well have determined his focus on the almost-sensuous beauty of Greek youths in the first place."

However influential Winckelmann's sexuality may have been in animating his aesthetic preoccupations and moral judgements, they found a large and appreciative audience. Winckelmann wrote about these sculptures as expressing an ideal beauty; he regarded them, Mosse tells us, in abstract terms. In fact, it was the absence of any individual or eccentric traits that defined the beauty of Greek sculpture. Such abstraction drained these Greek youths of any pornographic potential, rendering their nudity only aesthetic (Mosse 1996). This stripping of individuality for the sake of an abstract, presumably universal, beauty resonated, Mosse reports, with the notion, voiced by the English painter Sir Joshua Reynolds, namely that the ideal of beauty is constituted by its general principles, and is thereby superior to individual nature.

This preoccupation with masculine beauty must be set, Mosse argues, within a general understanding of how important a place the ideal of beauty occupied in an ever more visually oriented age. This ideal increasingly informed the symbols of private and public life. "Though the ancient Greeks had already privileged sight over other senses," Mosse (1996, 33–34) tells us, "and the Baroque had transformed the world into a stage, now the visual register was secularized and extended, becoming part of the rhythm of daily and political life." This was, Mosse reminds, daily life during the Industrial Revolution. The unification of the "beautiful, the true and the holy" functioned as a distraction from the ugliness of the modern age (Mosse 1996, 34).

Mosse reports that Winckelmann was praised as the re-discoverer of the "realm of beauty" at a university celebration in the mid-nineteenth century honoring his birthday. In the depths of the industrial revolution, when for many any sense of beauty in daily life was obscured by the smoke of factories and an intensifying commercialism, Winckelmann was singled out as the giver of light. Winckelmann's birthday was celebrated yearly in many German universities, and his conflation of beauty and manliness underwent a process of institutionalization that reached beyond the universities into the secondary schools (Mosse 1996).

Women were conspicuously absent from Winckelmann's principles of beauty, and women were conspicuously absent from what beauty symbolized for society's (male) self-image. Supreme beauty for Winckelmann was male, not female, and the examples of such beauty upon which he focused, such as the Apollo of Belvedere, were not only not women, they were not even androgynous. As Mosse (1996, 35) points, these young male nudes were "real men" because female influences were absent.

The power of those ancient Greek male nudes, sublimated in Winckelman's abstract formulations of beauty, stimulated others, among them Johann Gottfried von Herder, who saw Winckelmann as a Greek who had risen from the ashes of his forgotten people in order to illuminate his age. Goethe wrote copiously about Winckelmann, in whose aesthetic he discerned a legitimation of his ideal of the autonomous human being who must educate himself through art to a greater humanity. Winckelmann's ideals became embedded in the theory of *Bildung*, that middle-class conception of self-education and character building that in central Europe was imagined to create good citizens (see Westbury et al. 2000). Even critics admired Winckelmann, or, at least, his subject matter (Mosse 1996).

Winckelmann's influence extended well beyond Germany. His work was received enthusiastically in France as well. Not two years after he was translated in 1755, the *Encyclopedie* praised him for establishing that the ancient Greeks created the ideal beauty. His *History of Ancient Art* became a focus of the political as well as artistic debates taking place in France at that time. It is not surprising, Mosse notes, that Jacques-Louis David, an important painter during the French Revolution, very much admired Winckelmann. In England, a century later, Walter Pater was entranced by the elegance and balance of Winckemann's Greek sculptures. Robert Knox, the famous Scottish anatomist, saw in those naked Greek figures a mixture of robustness and vitality that he imagined the northern race had inherited (Mosse 1996).

Knox's racialization of male beauty also resonated with many. By the second half of the eighteenth century, Mosse tells us, a standard for male beauty had been established that stressed national or racial peculiarities, even among those who shared much the same ideal. The apparently ubiquitous interest in the naked Greek youth suggests that "race" and "nation" were not segregated or distinct concepts: both were laced with "homosexual desire," at least from the eighteenth century on. Mosse reminds us that German advocates of an "Aryan race"—introduced by Sir William Jones in the 1780s (see Wahrman 2004, 116)—believed that in their journey from their place of origin in the Far East, the Aryans had passed through Greece and incorporated the best of the ancient world, "the best" a code phrase for the appreciation of muscular young men.

While Winckelmann's ideal of male beauty was based on presumably colorless Greek sculptures, Carl Gustav Caro, in his *Symbolism of the Human Form* (1853), proclaimed that blond coloring derived from the sun was an additional mark of superior peoples, a curious "discovery" in light of the widespread belief that tropical sun turned the skin black, signaling inferiority. For Caro, the bodily structure of Greek sculpture was superior to all others, a bodily structure he saw also in young "Aryan" men. There were Nazis who saw themselves as "Aryan" expressions of this ancient Greek male ideal (Mosse 1996).

The European standard of male beauty had now been set; training had to be devised so that it could be attained. Gymnastics would become that training. Mosse notes that Winckelmann had written about the Greek gymnasium where exercises revealed the contoured muscled beauty of the naked male body, suggesting that "sport" and "working out" had explicit homoerotic elements from the very beginning. But it was not Winckelmann who popularized gymnastics; nor was it the widespread fascination with the (ancient) Greek male body. Rather, during the eighteenth century, gymnastics was seen as a means of personal hygiene. To illustrate, Mosse (1996, 40) quotes André David Tissot who wrote in 1780: "Gymnastics is that part of medicine which teaches maintenance or restoration of health by means of exercise." In his *The People's Handbook of Health* (1761), published around the same time that Winckelmann's most important works appeared, Tissot refers not to the ancient Greeks but to so-called primitive peoples who were, presumably, close to nature. These young (black?) men epitomized, presumably, wholesome and manly bodies.

"Sport" may have had its Western origins in homoerotic exhibitionism and competition (see Goldhill 1996, 19), but in the eighteenth century (as today) such exhibitionism was to be strictly disciplined. Gymnastics was said to control unlawful passions, substituting looking and being looked at for sexual encounter, a substitution that occurs in Cohen's interpretation of what happened inside Noah's tent. (No substitutions for Schreber: he felt God's desire on the surface of his body.) This presumably moral imperative became expressed discussions about physical exercise. The tendency to conflate body and soul, mentioned earlier, was fundamental to athletics as training for modern masculinity (Mosse 1996). Mosse points out that for Rousseau, bodily exercise was in no way subordinate to matters of the spirit: both were prerequisites for a life lived according to nature.

In this context, "nature" would seem to be a substitute for "culture," ancient Greek pederastic culture. In the context of early-twentieth-century America, "black" would seem to be a substitute for "Greek." The inner spirit, now surfaced in the outer form of the (male) body which all can see, becomes secured, anchored in the visible. Fixing the visual scene, evidently, destabilizes the position from which one sees. Substituting body for spirit, desire for identification and African men for ancient Greek boys proved disorienting for European men.

"Disorientation" in "positionality," Edelman (1994, 275) points out, follows from the threat associated with the "sodomitical" scene. Men looking at the naked bodies of other men, even when rationalized aesthetically and athletically, threatens to dissolves the homosocial structures sublimation solidifies. The indeterminacy and the disorientation of positionality that accompanies the breaking of the taboo of looking precipitated a gendered crisis of late-nineteenth-century European and American cultures, a crisis which threatened not only the gender order, but the racial one as well.

SODOMITICAL SPECTACLES

As Noah's terrible curse of his son belies and as the general biblical hysteria about homosexuality suggests, the son's desire for the father is also primary in biblical tradition. (Regina M. Schwartz 1997, 111)

Within a patriarchal culture, the more intense male homosocial desire becomes, the more intensely male homosexual desire becomes stigmatized and proscribed. (David Savran 1998, 186)

[M]arriage comes to rescue of besieged masculinity, making that masculinity whole again by redirecting its supposedly pathological sexual desires. (Mason Stokes 2001, 20)

[W]hite masculinity becomes what psychoanalysis calls a nodal point, an anchor in the constant slippage of meaning. (Thomas DiPiero 2002, 13)

Laboring to stabilize his patriarchal authority over his psychoanalytic sons, still struggling in the aftermath of his homoerotic relationship with Fleiss, Freud was living the relations among disorientation, positionality and sexuality. Lee Edelman (1994a, 275) ascribes Freud's inability to clarify the meaning of the primal scene to the destabilizing consequences of observing the sodomitical "spectacle," specifically, the unraveling of "positionality." Such a destabilization of positionality, Edelman notes, allowed Freud to identify with the infant whose spectatorial pleasure has now become internalized as aggressive, indeed, as a threatening self-scrutiny and potentially self-shattering guilt over vicarious anal eroticism. This shifting situatedness and the multiple identifications it invited disabled Freud from resolving the theoretical questions provoked by the primal scene, including the position he imagines the infant occupying within the scene itself (Edelman 1994a).

Edelman reminds us of the homophobic, homosocial, homoerotic, and homosexual relations that circulate within and structure the Western philosophical tradition. At issue for Derrida, Edelman points out, is the irreducibility of presence or absence, a binary logic that Derrida describes as intrinsic to "phallogocentrism," a "system of the symbolic, of castration, of the signifier, of the truth" (quoted in Edelman 1994a, 282). Derrida's philosophical performances of rigorously indeterminate situations recalls, for Edelman, Freud's inability to explain definitively the primal scene.

In his association of Derrida with Freud, Edelman links the (il)logic of paired opposites, the institution of sexual difference (via castration and the male fear thereof), and the developmental surpassing of a pregenital ambivalence, the multiplicity of identificatory positions that renders distinctions such as inside or outside, imagined or real, indeterminable and indeterminate. As the "opposite" of pregenital ambivalence, Edelman (1994a, 283, emphasis in original) theorizes, castration is the "*knowledge*" of "antithetical positioning," achieved through the "very *principle*" of "paired opposites." "Truth" becomes the "either/or" "determination" of "presence" or "absence" (Edelman 1994a, 283). Castration and, in the male, its defensive denial of the curse, creates, in this analysis, a series of binaries; not least among them are "opposite" sexes (even within the male "sex") and distinct "races." Not only incestuous desire gets restructured in a defensive cover-up of the "sodomitical scene," so does its supplement, "race."

These sodomitical dynamics of the primal scene invoke such a sense of catastrophe that only banishment, servitude, and enslavement can recast the scene, restructure it as patriarchal—indeed divine—authority and order. In these dynamics we discern the multiple births of "phallogocentric" philosophy, compulsory heterosexuality, and a sexualized racialization in which black bodies become substitutes (or projective screens; see Young Bruehl 1996) for banished and enslaved incestuous desire. Such "anarchy" must be restructured as "order": white men as "masters," black men and women as "slaves."

This arbitrary and brutal order was unstable from its inception, bound to collapse, if the instability of the father–son relation encoded in Genesis 9:24 is any indication. Discussing the (untranslated) work of French psychoanalyst Wladimir Granoff, Lukacher (1986, 1966) points out that "the father–son relation may have incorporated the terms of its own reversal: that is, Freud and the Wolf-Man as alternately father and son in relation to each other." In his introduction to the (also untranslated) work of Nicolas Abraham, Lacan takes note of the Wolf-Man's "negative" oedipal complex, in which the patient has become "hysterical insofar as he is disappointed not to have been seduced by the father" (quoted in Lukacher 1986, 159). Not to be disappointed, Schreber will be seduced.

Within the either/or logic that the heterosexually identified Freud practiced and that made possible the law of castration, there is, Edelman (1994a, 284) argues, a "sodomitical (il)logic" both "before" and "behind" the primal scene. In the politics of discursive practices, Edelman (1994a, 284) contin-ues, the representation of sodomy between men threatens the "epistemolog-ical security" of the "observer," for seeing the "sodomitical encounter" blurs "positional distinctions" as it implicates him in a "spectacle" that promises "castration." In his epistemological insecurity, does the observer then trans-pose revulsion into "yearning," as Bernard Yack (1996) characterizes it, "for total revolution"?

For Freud, then, primal fantasies required the notion of "origin," a fact that recalls the patriarchal obsession with genealogies. The "inferential" character of paternity in which, except by rather recent DNA tests, the father disappears or, at least, can go unrecognized, may have provoked this com-pensatory attribution of significance to "origins." For Freud, the primal scene is that construct the representation and "solution" of which constitutes a major psychological challenge for each European.[4] In the Wolf-Man's pri-mal scene, it was the radical indeterminacy of the patient's spectatorship that enabled his identification with his father and desire for his mother to merge with each other, or, simply, switch, creating, for him, an unsatisfactory solu-tion, a man whose anal eroticism and haunting dream of wolves drove him to Freud (Moglen 1997; Laplanche and Pontalis 1973). Are these the unnamed dynamics that brought Ham to his naked father's tent?

Freud was not the first therapist the Wolf-Man consulted. In his *Memoirs*, which he wrote in 1970–1971 at the age of eighty-three (Gardiner 1971), the Wolf-Man recalls the therapy he received in St. Petersburg in 1908 when, after his sister's suicide, he suffered acute depression and withdrew from the uni-versity there. His father had intervened, sending him to the city's leading neu-rological expert. In his *Memoirs*, the Wolf-Man refers to him only as Professor B., but to Obholzer (1982) he identifies him as Dr. W.M. Bechterev, a leading Russian psychiatrist and author of the book *Suggestion and Its Role in Society* (1898). Ned Lukacher (1986, 144) observes: "The patient seems to have been permanently caught somewhere between hypnosis and analysis." Is the latter a rational and transferential expression of the former?

In the sodomitical primal scene, the birth is "virgin," but what is missing is not the (God-the-) Father but the Mother (Mary). In this founding scene of "race" in the West, the mother's body is banished and "blacks" are born in an "unholy"—that is, desublimated and thereby abject—(de)coupling of father and son. The sexual coupling of father and son is cursed and becomes, in its political structuration, the enslavement of the black body which, in part because "its" nature is imagined as "radical indeterminacy," must be domi-nated. From the curse of Ham comes the curse of whiteness.

Whatever appears to the subject as the beginning—although, as Edelman persuades, part of the problem of the sodomitical scene is that temporal as well as positional logics become unstable and the switch–of his or her current "problem" constitutes the primal moment or original point of departure of a history. In the primal scene, Laplanche and Pontalis (1973) explain, it is the origin of the subject that is represented; in seduction fantasies, it is the origin or emergence of sexuality; in castration fantasies, the origin of distinction between the sexes. In Genesis 9:24, these fantasies merge, as the origin of the "black man" occurs after the son's seduction of his father (or is it the father's seduction of the son?), rendering the father castrated (or covering up), a wound which must be disguised by the (re)assertion of patriarchal authority, in which the son is disclaimed, his progeny enslaved, castrated, or emasculated in civic terms. The "wound" circulates from father to son, from Judea to Europe, from Africa to the Americas. The origin of the distinction between the sexes is the distinction between "men" and "sodomites" (after all, Eve is Adam's rib), and it becomes the distinction between men and women, between masters and slaves, between Europeans and Africans.

Although "race" may have been born, in the (white) Western mind, in gender, it cannot be subsumed in sex, even of the sodomitical kind. The bastard child becomes adult; race detaches from gender. Only in regressive analysis does its genesis become clear. Kaja Silverman—whose work is a prerequisite for any work toward "re-envisioning" the primal scene (Edelman, 1994a, 285 n. 3)—understands: "the unconscious articulation of racial and class difference is facilitated, however, by the articulation of an even more inaugural difference, which we also need to conceptualize ideologically—sexual difference" (Silverman 1992, 23). That is sexual difference between men, projected onto—demanded from—women, recast as racial difference. Kalpana Seshadri-Crooks (1998) does not disagree with Silverman that the family is a site of psychic reality, but she does question that the family is a "much more" significant element of psychic reality than, say, society or "race." To "understand" the "reality" of race, writes Seshadri-Crooks (1998, 356–357), we must confer upon it "coevality" with sex; failure to do so "trivializes" the "effects" of "racial identification." This is necessary, Seshadri-Crooks (1998, 357) argues, because while gender's "essential" meanings seem contested in "everyday life," the reality of race seems naturalized: there are "no challenges" to race beyond the "empty" academic acknowledgment that it is a "construct." She suggests that this resistance to race as coeval with sex is a consequence of the hegemony of "whiteness" as an ideological structure.

Whiteness does make itself invisible to itself, but it is a mistake, as is clear by now, to rule out a sexual (and, more broadly, gendered) genesis for whites' constructions of "blackness" and "race." I do not see why a sexual theory of race "trivializes" the effects of racial identification, or, pedagogically speaking, why, in fact, this might not be a good thing. I agree it is not useful to declare the family the fundamental site—the primal scene—of the human subject precisely because the "family" is not only fundamental, and even when it is, it is so because it funnels, intensifies, and singularizes society, the racial order, the sexual regime. But to focus on the sexualized character of "race" in whites, I suggest, is a way of paying respect to, not a way of trivializing, those whose lives were and are overdetermined by its effects, including racial identifications.

By engaging in a cultural psychoanalysis of "race" I am working to disassemble the convoluted genealogy of racial identification in whites in order to precipitate the racialized and gendered "self-shattering" of whiteness. I am going beyond the tip of the rhetorical white hat by working to disrobe its sedimented genealogies in the Western (white) imagination, the canonical documentation of which is the Bible. As in the Lacanian conception of analysis, my aspiration is to encourage the (especially white male) reader to "own" his or her "alienation" and "desire," by "confronting" him with "his own unconscious fantasy" (Newman 2004, 307). It is an incestuous fantasy.

Freud's construction of the primal scene, Lukacher (1986, 44) points out, represents an effort to define the "work of reparation in terms of the affirmation of the ineluctability of difference and deferral." Race is the deferred and displaced difference between Adam and (St)Eve, sexual difference among men, enacted in Noah's tent. This construction of the primal scene of race in the West is, as Lukacher writes in a different context, an "undecidable intertextual event that is situated in the differentiated space between historical memory and imaginative construction, between archival verification and interpretative free play" (1986, 24). Through "language's irrepressible effect of displacement" we can carve a psychic "opening," what Freud calls the "cure" (Lukacher 1986, 44).

How Shall I Know Thee?

[W]e need to divest ourselves from over-identifying with the victim on the cross. (Robin Hawley Gorsline 1996, 138)

Yahweh is linked to the light. (Lewis R. Gordon 1996, 244)

[P]enetration signifies domination and feminization. (Ann Cvetkovich 2003, 61)

Daniel Paul Schreber was, Louis Sass (1994, 7) points out, a "highly intelligent and articulate" man whose symptoms involved elaborate delusions—"so-called delusions," Schreber insisted—elaborated in his *Memoirs* (1903). There we learn that he was transgendered into a woman, able to discern the spheres of "nerves," "rays," "souls," and "Gods" all in constant contact with one another and/or with himself. These "supernatural matters" were, Schreber (2000 [1903], 44) wrote, the "most difficult subject ever to exercise the human mind." Not immodest, he cannot count upon being "fully understood" because his experience cannot be adequately expressed in "human language," exceeding "human understanding. "[M]uch remains only presumption and probability" (2000 [1903], 16).

As we have seen, Schreber's book provided the material for the only case study Freud ever wrote of a psychotic patient, his 1911 "Psychoanalytic Notes Upon an Autobiographical Account of a Case of Paranoia." Why did Schreber matter to Freud so much that Freud would base a major study upon his *Memoirs*? Santner (1996, 19) argues that the answer has to do with Freud's own conflicts over what he regarded (projected?) as Schreber's core issue: homosexuality. To document his claim, Santner points to Freud's letters written around the time of the composition of the Schreber essay, which suggest that Freud was still very much working to conclude emotionally his "homosexually charged relation" (Santner 1996, 19) with Wilhelm Fliess. Freud may have ended things with Fleiss, but his own homosexuality, Santner suggests, resurfaced in his relationships with various members of his inner circle.

To illustrate, Santner points to an October 6, 1910 letter Freud wrote to Ferenczi, who had accompanied him to Italy the previous summer, a period in which he was studying Schreber's *Memoirs*. In the letter Freud reports that working on the Schreber material had helped him overcome much of his own homosexual inclinations: "since the case of Fleiss, with whose overcoming you just me occupied, this need has died out in me. A piece of homosexual charge has been withdrawn and utilized for the enlargement of my own ego. I have succeeded where the paranoiac fails" (quoted in Santner 1996, 19). The conclusion of Freud's study of Schreber's *Memoirs* was his claim that paranoia represents repressed homosexuality.

Several months later, in letter to Ferenczi, Freud wrote: "Fliess—you were so curious about that—I have overcome. [Alfred] Adler is a little Fliess just as paranoid. Stekel, as appendix to him, is at least named Wilhelm." In an earlier letter to Jung, Freud claimed: "My erstwhile friend Fliess developed a beautiful paranoia after he had disposed of his inclination, certainly not slight, toward me." In another letter to Jung written during the Italian journey with Ferenczi (while working on Schreber), Freud characterized his traveling companion as feminized: "He has let everything be done for him like a woman, and my homosexuality after all does not go far enough to accept him as one" (quoted in Santner 1996, 20). Calling it "fairly typical in the literature on Freud's Schreber essay," Santner (1996, 20) quotes Peter Gay's suggestion that Freud positioned himself as, at least in some sexual sense, parallel to Schreber:

> Freud's rather manic preoccupation with Schreber hints at some hidden interest driving him on: Fliess. To study Schreber was to remember Fliess, but to remember Fliess was also to understand Schreber. Freud used the Schreber case to replay and work through what he called (in friendly deference to Jung, who had invented the term) his "complexes." (Gay 1988, 279; quoted in Santner 1996, 20)

Santner focuses not on Fliess but upon Freud, and, in particular, on his "surprising protestation" (Santner 1996, 20) concerning the originality of his conclusion on paranoia, derived from Schreber. Freud insists "that I had developed my theory of paranoia *before* I became acquainted with the contents of Schreber's book" (Freud, 1953–1974, 79; quoted in Santner 1996, 20–21, Santner's emphasis).

What to make of this "masculine protest," Santner (1996, 21) asks, borrowing the phrase from Adler, a phrase Freud too had used in his reading of Schreber. What strikes Santner most about Freud's eagerness to point out that his views on paranoia are not derived from Schreber's is that this "anxiety" is "uncannily reminiscent" of Schreber's "confusion" over the "originality" of "his own thoughts, thought processes, and language" (Santner 1996, 21).

Not unlike Schreber's experience of the malevolent forces assaulting his soul and body, Freud worried about the integrity of his body of knowledge: psychoanalysis. Both men, Santner (1996, 21) writes, are concerned they are only "parroting back" language "originating elsewhere." If Freud has a transference relationship with Schreber (through his memoirs), Santner (1996, 21) continues, its content is not only homosexual desire: it has to do with issues of "originality" and "influence" and "authority" in the very therapeutic movement Freud claimed as "his own." Do the two—anxiety over homosexual desire and questions of origin and influence—always operate together?

Such issues of "(be)hindsight" Lee Edelman discusses in terms of the notion of the primal scene, as discussed earlier in the chapter. For Freud, the primal scene has to do with the paternity of psychoanalysis. As psychoanalytic theorists subsequent to him would argue, it is the inferential character of paternity that animates men's historic interest in patrilineage which is, in one sense, a birth copyright, claiming property (Chodorow 1978). Why should Freud—anxious already over his ethnicity (see Gilman 1993) and sexuality—be exempt?

Santner provides another instance of Freud's anxiety over issues of authority and influence. This moment, Santner argues, comes just after Freud suggests that Schreber's second illness was precipitated by a homosexually charged longing for his psychiatrist, Paul Flechsig. Freud writes that "this feminine phantasy, which was still kept impersonal, was met at once by an indignant repudiation—a true 'masculine protest,' to use Adler's expression," but, Freud points out, "in a sense different from his" (Freud, 1953–1974, 42; quoted in Santner 1996, 24). Freud provides a footnote in which he adds: "According to Adler the masculine protest has a share in the production of the symptom, whereas in the present instance the patient is protesting against a symptom that is already fully fledged" (Freud, 1953–1974, 42; quoted in Santner 1996, 24).

Santner points out that when Adler presented his view to the members of the Psychoanalytic Society in January and February 1911, Freud had become distressed over paternity and property, claiming that several of Adler's key ideas, including that of "masculine protest," were, in fact, formulations taken from Freud's own prior insights. There were—here Santner again quotes Peter Gay—of "spurious, manufactured originality" (Gay 1988, 221–222; quoted in Santner 1996, 24). In his footnote to this quotation, Santner notes that Jacques Le Rider (1993) organizes his study of the fin-de-siècle crisis of gender, national and ethnic identity around this notion of "masculine protest," a topic also theorized by Gerald Izenberg, introduced in the first section of this chapter.

Freud's anxiety over originality and influence, Santner points out, was lived out during the early and crucial years of the psychoanalytic movement, a period riddled with increasingly divisive internal tensions. The final break with Adler would come in 1911, the break with Jung two years later. These events, Santner notes, intensified and complicated Freud's continuing efforts to gain recognition from the larger scientific and intellectual community.

Sander Gilman notes that Freud read Schreber's book in the midst of his confrontation with Adler, a simultaneity of events, Gilman (1993, 142) suggests, that re-stimulated Freud's "homoerotic identification with Fliess." Indeed, Freud wrote to Jung at the end of December 1910, telling him that the reason the breakdown in his relationship with Adler "upsets me so is that it has opened up the wounds of the Fliess affair." Like a letter composed the same month to Ferenczi, Freud told Jung that "my Schreber is finished" and that he was unable to "judge its objective worth as was possible with my earlier papers, because in working on it I have had to fight off my complexes within myself (Fliess)" (quoted passages in Gilman 1993, 142).

For Gilman, Freud's characterization of Schreber's illness in terms of repressed homosexual desire was not only associated with his experience of Fliess; it derived, as well, from Freud's conflict with Adler. Both Adler and Fliess were Jews, and Freud, Gilman underscores, worried that psychoanalysis would be dismissed as a Jewish undertaking. Due to this concern, Gilman suggests, Freud wanted to replace Adler with Jung. Schreber's *Memoirs*, Gilman argues, the very text in which Jung had been interested and had recommended that Freud read, was "one structured by the central metaphor of the dangerousness of the Jews" (Gilman 1993, 142). The association of Jung and Schreber's text enabled Freud to distance himself from the Jewishness and homoeroticism associated with Fliess and Adler (Gilman 1993).

For Gilman (1993), Freud's characterization of Schreber's illness as associated with Schreber's fear of or desire for castration and degradation enabled Freud to ignore the religious coloration of Schreber's system, something Freud regarded as an incidental rather than a primary aspect. Schreber's anxiety in "being simultaneously transformed into a woman and a Jew," Gilman (1993, 145) notes, "was ignored." Indeed, Gilman (see 1993, 154) points out that Freud never attended to this anti-Semitic aspect, an aspect Gilman (1993, 143) judges to be a "powerful subtext" in Schreber. Freud made no mention of the persistent incorporation of the rhetoric of anti-Semitism. Yet, up to this point in his reading of the *Memoirs*, Gilman notes, Freud underlined each reference to castration in Schreber's account, such as that on page 4 of the German text, which Freud annotated with the comment: "fantasy of feminization."

"The evocation of castration and its association in Freud's works with circumcision provide a context for Freud's inability to read aspects of Schreber's text," Gilman (1993, 145) argues. Gilman notes that others (among them William Niederland, Morton Schatzman, and Jeffrey Mason) have pointed out that Flechsig had proposed (and had used) castration to treat hysteria, and that Freud may well have been aware of this. But Gilman is less interested in the oedipal resonance of this possibility (although, he points out, it is not known if Schreber was also aware of his doctor's recommendations in cases of hysteria), than in its anti-Semitic resonance. "What is clear," Gilman (1993, 145) notes, "is that castration was understood at the time as an alteration of the body that made it different, and more feminine in that it also made it more 'Jewish.' " Circumcision, too, made the body more Jewish and, as

Bruno Bettelheim suggests, more feminine, all the while appearing to demarcate "manhood." In twentieth-century America, Michael Kimmel (1996) suggests, Jewish men were feminized because Jewish religious culture stressed morality and literacy; they were seen by many Christian and secular men as bookish, even effete. Kimmel recalls marching in a protest against the war in Vietnam when a heckler screamed at me to "go back to Russia, you Commie Jew faggot!" (quoted in Kimmel 1996, 277). Kimmel (1996, 277) writes: Though I was startled at the time by the venom of his accusations, stung by his rage, what is most significant to me now is the way that communism, Judaism, and homosexuality were so easily linked in his mind. All three, I came to understand, were not "real men." Let us praise men who are "not real."

CHAPTER 6

An Epistemology of the Body

THE GENDER OF KNOWLEDGE

I would like to meet the man who, faced with the choice of either becoming a demented human being in male habitus or a spirited woman, would not prefer the latter. (Daniel Paul Schreber 2000 [1903], 164–165)

It [Schreber's feminization] is rather a way of structuring a relation to God's *desire*, to the revelation that this ultimate master's knowledge and powers are *lacking*. (Eric L. Santner 1996, 99)

Although representing what is most emphatically our own, the language of our desire consequently remains for most of us irreducibly Other. In a certain sense, we do not even speak it; rather, it speaks us. (Kaja Silverman 2000, 51)

Vision as an "autonomous" process or exclusively optical experience becomes an improbable fiction. (Jonathan Crary 1999, 352)

For Louis Sass (1994, 156 n. 45), God's interest in Schreber amounts to a "pantoptical arrangement" that "makes the individual feel constantly exposed to an external, normalizing gaze, thus subjecting him or her to the dictates of an authority that must ultimately be internalized." Sass (1994, 158) cites "recent historical research" suggesting that God-the-Father is no simple substitute for the family father: such research suggests that Schreber's upbringing may not have been atypical for his time and place. Rather than only or, perhaps, even primarily a consequence of his particular family, then, Schreber's illness, Sass (1994, 158) offers, may indicate "how schizophrenic symptoms can develop within the (panoptical) modern social order." Sass is thinking of Foucault here, specifically of his *The Order of Things* and *Discipline and Punish*.

Such internalized self-monitoring resides, presumably, in the head, in reason. To quiet what Schreber called "compulsive thinking" and his self-conscious "nerves of intellect," Schreber succumbs to physical pleasure, for him associated with femininity. "The experience of feminization was," Sass (1994, 126) suggests, "Schreber's major antidote to the intellect, his palliative for the self-torturing mind." For Schreber, "feminization" and "femininity" had inherently little to do with concrete women, but seem, instead, to be split-off fragments of himself. For Schreber, feminization may have been a marker of sexual persecution and gendered defeat but, as Sass points out, it had soothing and reassuring effects, calming the restlessness and self-assault precipitated by his compulsive thinking and hyper–self-consciousness, two cultural characteristics of whiteness.

Schreber became convinced that "the feeling of sensual pleasure— whatever its physiological basis—occurs in the female to a higher degree than in the male, and . . . the *mammae* particularly play a very large part" 1968 [1903], 205; quoted in Sass 1994, 126). It was the so-called passive experience of pleasure that Schreber associated with femininity (Sass 1994, 126). In order to feel "feminine," Schreber would stroke himself or, rather, was stroked, by his heavenly Father. Unlike the heterosexual male masturbator— who, presumably, maintains an "active" sense of his phallic potency even though he is a man acting upon himself, i.e. homosexually—Sass notes that Schreber felt identified not with the active (masculine hand), but with the "passive" (in his mind, the feminine). He became the flesh that was being touched, the skin being stroked, the soul-body being taken. By stroking himself Schreber experienced what he took to be the female "nerves of voluptuousness" that he imagined layered beneath the skin of women:

> When I exert light pressure with my hand on any part of my body I can *feel* certain string or cord-like structures under the skin . . . . Through pressure on such structures I can produce a feeling of female sensuous pleasure . . . . (1968 [1903], 205; quoted in Sass 1994, 126–127)

Such imagery suggests that underneath the masculine identification accomplished through socially learning—under threat of castration—is that preoedipal symbiosis with the mother. Before the Word was the (M)Other.

As Sass points out, Schreber insists that "I do this, by the way, not for sensual lust, but I am absolutely compelled to do so if I want to achieve sleep or protect myself against otherwise almost unbearable pain." This experience of female "voluptuousness" is associated with what he calls "Blessedness," a state free of the divisions of hyperconsciousness and therefore infused with the sentiment of being (1968 [1903], 111; quoted in Sass 1994, 127). For Schreber, as for the West (see, for instance, Bordo 1993), femininity represents the sensual, a state of "Blessedness" in which the insidious anxieties of "compulsive thinking" are soothed. Does the repudiation of the son's maternal identification, reiterated in the father's repudiation of the son's sexual desire for him, portend "compulsive thinking" abstracted from the sensual, from the body, from the earth, for example, whiteness? Did a Christian culture of self-torture and self-mutilation portend the torture of others?

Touching himself was not, Sass points out, Schreber's only means of engendering his own feminization. As he lay in bed caressing himself, Schreber sometimes imagined himself standing before a mirror gazing at his own, now feminized, body. Is this the gender of the mirror stage, prompting boys to misrecognize themselves as not-girls, that is, prompting the denial of (what they imagine as) lack? The feminization that Schreber experienced before the mirror (whether in his imagination or in fact before the glass) was, Sass (1994, 127) suggests, also an "epistemological" feminization, in which femininity was not only a lived and kinesthetic bodily presence but, as well, a kind of "epistemic objecthood." Schreber comments: "I consider it my right and in a certain sense my duty to cultivate feminine feelings," and "mere low sensuousness can therefore not be considered a motive in my case" (1968 [1903], 207–208; quoted in Sass 1994, 127). It is to experience the "solidity" of "objecthood," Sass (1994, 127) speculates, that ontological grounding Sartre terms the *en-soi* or "in-itself," that also moves Schreber. Sass (1994, 127–128) writes: "Feminization of this sort implies giving up aspirations toward being a constituting center and choosing instead the rather different form of power inherent in the role of attracting and fixing the attention of an other."

As Sass (1994, 118) observes, Freud decoded Schreber's psychosis as repressed homosexuality, a "formation . . . that both expresses and disguises an underlying homosexual fantasy rooted in a preoedipal psychosexual fixation." Because, Freud reasoned, Schreber cannot bear to allow himself to experience desire for a man, the desire is first reversed, then projected. In Schreber's case, desire for a man—the father—becomes hatred for the father, which, in turn, became God-the-Father hates me. It was a similar sequence for nineteenth-century Southern white men vis-à-vis black men, a desire also rooted, I reasoned, in the preoedipal phase, regression to which had been precipitated by the gender-shattering experience of defeat in the Civil War.

Sass (1994, 119) reviews the major "revisionist" interpretations of Schreber's case, noting that these—most prominently that of Elias Canetti— emphasize not sexuality but power (see, also, Santner 1996, ix). For Canetti, in fact, Schreber's *Memoirs* foreshadows Hitler's *Mein Kampf.* Eric Santner (1996, x) suggests that although "far more sympathetic to the ambiguously transgressive dimensions of Schreber's delusions," Gilles Deleuze and Félix Guattari "second" Canetti's reading. "[S]uffice it to say," Santner (1996, x) concludes, that these studies of Schreber establish a "powerful link" between the *Memoirs* and the "core features" and "obsessions" of National Socialism. How we wish that Schreber the "girlie man," not the "muscled" Austrian, had prevailed!

Other versions of this "power" interpretation—Sass names those of William Niederland (in 1953) and Morton Schatzman (in the 1970s)— emphasize the authoritarian child-rearing practices of Schreber's father. As Sass reads these, each ascribes Schreber's delusions to his father's authoritarian child-rearing practices, designed to suppress much of the child's natural spontaneity, willfulness, and independence. In these revisionist readings, the sexual themes—especially Schreber's transformation into a woman—are less a function of libidinal wishes than they are of power relationships. It is not clear to me how the distinction between power and sexuality can be drawn so sharply (see, for instance, Brownmiller (1993 [1975]).

Although acknowledging that these "interpretations" provide "valid insights," Sass (1994, 119) regards them as "incomplete" and "superficial." In his view, to understand the meaning of sex and power in Schreber's world, "one must first understand a dimension that is more fundamental for such individuals, that of knowledge." What Sass proposes, then, is that we consider the issues of power and sexuality in Schreber's world in terms of the epistemological themes. Why "knowledge" as a category is more fundamental than "gender," Sass fails to explain. In Schreber's case at least, I submit that the two cannot be understood apart from each other.

Sass (see 1994, 120) asserts that power is not the issue for Schreber but, rather, knowledge, pointing out that what Schreber feared at the onset of his illness was not that someone was trying to control his behavior but that there was a conspiracy to "destroy his reason" (1968 [1903], 211–212), his capacity to be a conscious center in Sass' terms. For Sass, Schreber's experience resembles the philosophical doctrine of solipsism, "according to which the whole of reality, including the external world and other persons, is but a representation appearing to a single, individual self, namely, the self of the philosopher holding the doctrine" (Sass 1994, 8). "Solipsism was a recurrent, perhaps even an obsessional concern," Sass (1994, 8) notes, of the philosophical work of Ludwig Wittgenstein, famous for his accusation that much of traditional philosophy, with its obsession with metaphysical speculation, amounted to a kind of disease—even to a mental illness. For Wittgenstein, solipsism was a primary symptom of this metaphysical or philosophical disease, "a disease born not of ignorance or carelessness but of abstraction, self-consciousness, and disengagement from practical and social activity" (Sass 1994, 9).

Sass (1994, 121) is not just saying that power and sexuality exhibit a "distinctively epistemological cast" in Schreber's experience. Sass is saying that both power and sexuality in Schreber can be understood only in the context of an epistemological crisis. For Schreber, the distinction between masculine and feminine is a distinction between mind and body, subjecthood and objecthood, which are, he underscores, "epistemological" and "ontological modes of being" (1994, 122). Precisely for this reason, I would add, "knowledge" cannot be split off and made more "foundational" than gender and power; ontology structures epistemology: its structures and processes are revealed epistemologically.

For Eric Santner, Schreber's experience—and Freud's interpretation of it—require us to focus on the historical moment. Both Schreber's *Memoirs* and Freud's theorization of them were written and published during the early moments in the institution of psychoanalysis. Santner points out that this period of institution was one in which psychoanalysis was (as we have seen) fragile, vulnerable, a time when Freud's work was under attack. For Santner, this crisis of symbolic authority provides the necessary background for making intelligible Freud's passion for the Schreber material. Freud's "preoccupation" with "originality"—intellectual and institutional paternity, I might add—suggests, Santner (1996, 25) offers, his "profound" and "defensive" sensitivity to the "performative force" of his colleagues' "contestation" of the fundamental concepts of psychoanalysis.

It would seem that it is the fragility (inferentiality?) of paternity that may have empassioned Freud. Not incidentally, the fragility of paternity (originality in Santner's terms) turns on the "crucial features" of Schreber's breakdown, including the "central fantasy" of "feminization" (1996, 24). This is, I suggest, not only Schreber's idiosyncratic fantasy or Freud's historically contingent compulsion, but, rather, a central and enduring fantasy of Judaic-Christian civilization that, by the late nineteenth century, had become distinctively racialized. It was first a gendered crisis, following from the ancient-Israelite (male) anxiety of being erotically engaged with a male God, the cultural precondition for the West's "peculiar institution" of racialized enslavement. It was, presumably, a servitude ordained by God in which power and sexuality are inextricably intertwined.

RESTRUCTURING THE MALE BODY

[A]nd there be eunuchs, which have made themselves eunuchs for the kingdom of heaven's sake. He that is able to receive *it*, let him receive *it*. (Matthew 19:12)

[M]en may meet God only as women. And circumcision makes them desirable women. (Howard Eilberg-Schwartz 1994, 174)

To think whiteness requires us to rethink both racism and the ways racism speaks the body. (Phil Cohen 1997, 246).

The various orthopedic devices Schreber's father employed for the correction of children's posture and to prevent their masturbation may well have contributed to Schreber's castration. "They are machines," Gilman (1993, 160) suggests, "for the restructuring of the body, they are machines that feminize Schreber's body by unmanning him with magical rays. These machines also make the body into a Jew." Schreber focuses on his feminization and the cross-dressing that followed, allowing that others may find it "somewhat unreasonable" that "at times I was seen standing in front of the mirror or elsewhere with some female adornments (ribbons, trumpery necklaces, and suchlike), with the upper half of my body exposed" (1968 [1903], 300; quoted in Santner 1996, 81). Schreber defends his feminization by arguing that he is very careful to engage in cross-dressing only when he is alone. He further notes that since his female adornments are for the most part inexpensive, his transvestism cannot be cited as an example of poor judgement in the management of his financial affairs.

Schreber's primary defense of his cross-dressing is, however, that it is a religious practice, one necessitated by the special relation he has come to have with God. Regarding his delusional system in general, Schreber asserts: "I could even say with Jesus Christ: 'My Kingdom is not of this world'," pointing out that his "so-called delusions" were concerned "solely" with God; "never" did they "influence" his behavior in any "worldly matter" (1968

[1903], 301; quoted in Santner 1996, 82). He then adds: "apart from the whim already mentioned [i.e. cross-dressing], *which is also meant to impress God*" (1968 [1903], 301–302; Santner's emphasis). Like ancient-Israelite men, like Protestant brides of Christ, this son submits, not by sublimating the incestuous bond of father and son, but by enacting it.

Gilman (1993, 162) suggests there is a "general parallel" often drawn between the Jew and the homosexual in many writings of assimilated Jews, Jews who worried they were perceived by the Christian majority as deviations from the norm. The future foreign minister of the Weimar Republic, Walter Rathenau, begins a 1897 essay by "coming out of the closet" as a Jew, nonetheless condemning Jews as a foreign body in the cultural and political world of Germany (Gilman 1993). Zionism and psychoanalysis were, Daniel Boyarin (see 1999, 38, 222) proposes, "cultural answers" to this gendered political problem faced by Jews at the end of the nineteenth century in Europe.

That Schreber can experience his homosexual desire only as a woman (and "woman" as archetype or sexual object), and, therefore, as "castrated" by God-the-Father, suggests that he experienced homosexuality in heterosexual terms. Such an organization of desire was completely unacceptable to Guy Hocquenghem, a key theorist of the 1970s whose work anticipated queer theory (see Hocquenghem 1978; Marshall, 1997). Like Deleuze and Guattari, Hocquenghem regarded Schreber's illness not as paranoia but schizophrenia, a radical decoding of heteronormativity. "The very notion of object-choice," Marshall (1997, 32) tells us, "presupposes a bodily unity integrating the drives, and the entry into a system of binaries, of the similar and the different." Such unity could be imposed only *retrospectively*, leaving residues of longing, experienced, depending upon the object chosen, as homosociality or opposite-sexed emotional intimacy. Marshall continues (1997, 32):

> Homosexual object-choice is thus a matter of entering a binary system suffused with guilt because homosexual *desire* is otherwise sublimated and is a source of social anxiety, or to put it another way (quoting Hocquenghem) "sublimation is simply homosexuality in its historical family truth."

As such, homosexual desire constitutes "a crack in the system," destabilizing normativity, recalling an earlier polymorphous perversity. It is a polymorphous "order" in which, among other things, God is not all-powerful, but, in fact, all-too-human.

Acknowledging Freud's decoding of Schreber's conception of God's *lack* of omnipotence as an oedipal assault on God and on the paternal authority generally, Santner (1996, 61) reminds us that Schreber himself was convinced that this "lack" constituted the so-called Order of the World. This presumably cosmic law was transgressed by excessive and prolonged "nerve contact" between God and Schreber, an order of bodily penetration of the son by the Father. The "Order of the World" functioned to protect not only human beings from the overwhelming presence of God, but to protect God himself from the dangers of excessive nerve contact. God had to protect himself? Schreber anticipated the question, explaining that the "nerves" of "*living* human beings," especially during states of "*high-grade excitation*," hold such "attraction" for God that "He" would be unable to "free Himself" from them, thereby "endanger[ing] his own existence" (1968 [1903], 48; quoted in Santner 1996, 61; emphasis in original). The Father sexually stimulated by his son is no longer God.

Schreber's insight into God's internal division reveals (characteristic of the illness, Freud believed) a tendency toward splitting. The father figure, Santner suggests, is divided into two distinct paternal positions. The first is distant, characterized not by omniscience, but, rather, by a "peculiar ignorance" (1996, 62) about human life, including bodily functions. Such divine ignorance, Schreber emphasizes, accords with the law. In the second paternal position, ignorance becomes knowledge, distance becomes proximity, and God-the-Father becomes "obscenely" (1996, 62) engaged in human life, enjoying men's sexual pleasures, their most secret thoughts and dreams, even bowel movements, flows affected, recall, in the case of the Wolf-Man. The entire "plot" of the *Memoirs*, Santner (1996, 62) offers, reports Schreber's "struggle" to "integrate" these two fathers, to "reconcile" the "extralegal paternal presence" with the Father identified with the "Order of the World" and the "law of proper distances." It is precisely "the law of proper distances" that disappeared inside Noah's tent and became institutionalized in racialized enslavement.

The Schreber case is not the only occasion on which Freud theorized the homoerotic relations between fathers and sons, Eilberg-Schwartz notes. In his analysis of the "Wolf Man," Freud also discusses how the worship of God expresses homoerotic attachments to the father that, in concrete relationships between fathers and sons, must be suppressed. In the "Wolf Man" case, this is evidenced in the young boy's questions concerning Christ's "behind" and his capacity to excrete "waste." Here "we catch a glimpse of his repressed homosexual attitude in his doubting whether Christ could have a behind, for these ruminations could have no other meaning but the question whether he himself could be used by his father like a woman—like his mother in the primal scene" (Freud, 1918 [1914], 64; quoted in Eilberg-Schwartz 1994, 37).

Because God-the-Father is a "spiritual projection," Eilberg-Schwartz (1994, 37) summarizes, loving him is "less threatening" than loving a living father of "flesh" and "blood." The homoeroticism inherent in the father–son relationship is "transferred" to an "idealized" and "disembodied" Father, thereby sublimating the son's "homoerotic attachment" through its expression in an "idealized love" (1994, 37). To put the matter more bluntly: the body of God is banished so that it does not make explicit the homoerotic character of men's worship of him. In this view, God turned his back to Moses to prevent the Israelite men from facing the fact that the sacred is also the sexual. Because God's back was turned, Israelite men did not have to confront the male body, specifically the phallus, of the Father God they worshipped (Eilberg-Schwartz 1994). "From behind," Hocquenghem (1978, 87) quipped, "we are all women."

It was Moses, Eilberg-Schwartz reminds, who institutes the renunciation of (homosexual) desire through the prohibition on images, although the precedent had been established in Genesis 9:23. The prohibition institutes, through disavowal, scopophila, that sexual pleasure in looking so prominent in racialization. The prohibition against seeing the father's body leads, presumably, to a renunciation of a desire for the father's body. Such homoeroticism between father and son may have informed, Eilberg-Schwartz suggests, Freud's writing of *Moses and Monotheism*. In asserting the analogy between monotheism and homosexual latency, Freud cites a case to illustrate his point. Significantly, Eilberg-Schwartz observes, it is a case that involves the "negative" oedipal complex of a boy who watched his parents having sex. At first, the son's response was aggressively masculine, identifying with his father and expressing sexual interest in his mother. This pattern persisted until his mother presumably forbade him to touch his penis and threatened to inform his father (Eilberg-Schwartz 1994).

It is a sodomitical spectacle, as Edelman makes clear. Freud writes: "Instead of identifying himself with his father, he was afraid of him, adopted a passive attitude to him and, by occasional naughtiness, provoked him into administering corporeal punishment; this had a sexual meaning for him, so that he was able to identify himself with his ill-treated mother" (quoted in Eilberg-Schwartz 1994, 56). Although Freud does not make the connection, Eilberg-Schwartz does: the boy's renunciation of his (homo)sexuality is equivalent, in monotheistic terms, to the prohibition of images. After identifying with his mother and becoming "passive" vis-à-vis his father, the son loses interest in his penis after assuming a passive role to his father. Is this an allegorical restatement of the evolution of the white man, that is, that constellation of psychological repressions and desires that becomes the cultural matrix "he" takes with him when he invades Africa in the sixteenth century and there sees the bodies of black men, men who, it seems to the European, have not renounced the body? By enslaving Africans (overwhelmingly men), does he re-enact his own self-torture, his own cultural and religious enslavement?

We are forced to acknowledge the reciprocal relations between sex and the sacred, between repressed homosexual desire and patriarchal privilege, between the worship of a invisible God (through his scantily clad son hanging limp on a cross) and disavowed homosexual incestuous desire. Recall, as Eilberg-Schwartz asks us to do, Freud's claim that the prohibition on images prompts the discovery of paternity. Both, Freud suggests, represent the triumph of the spirit or intellect over sensuality. The prohibition on images forced people—specifically men—to experience God subjectively rather than imagine him visually, with a (male) body. Freud posits paternity as based on reason and not the senses, a logic structuring the Chain of Being during the Age of Reason.

Disavowal Produces Projection

In projection, dangers are inner dangers disavowed. (Joseph H. Smith 2004, 352)

Significantly, projection is a form of defense closely related to disavowal. (Kaja Silverman 1988, 16)

What happens to the foreclosed visual material thrown out of the psyche? (Martin Jay 1993, 585)

Freud, Santner points out, focuses on Schreber's earthly substitute father, namely Flechsig, whose relationship to Schreber led Freud to postulate the notion of transference. After his breakdown Schreber was in a state of extreme vulnerability, in a regressed and infantilized state in which he could experience once again the negative oedipal complex. Like the father, the good doctor promises the boy the "real" world, specifically property and power. Freud notices that Schreber's worldly success lacks completion, namely the reproduction of his patriarchy:

> His marriage, which he describes as being in other respects a happy one, brought him no children; and in particular it brought him no son who might have consoled him for the loss of his father and brother and upon whom he might have drained off his unsatisfied homosexual affections. His family line threatened to die out, and it seems that he felt no little pride in his birth and lineage. (Freud, 1953–1974, 57–58; quoted in Santner 1996, 49)

Drained off?

It is, rather, the first clause of the last sentence that Santner (1996, 49) finds as one of the more "telling formulations" in Freud's essay, pointing to a "crisis" of "symbolic function," manifested sexually. Santner (see 1996, 49) links the two. I point to the last clause of the first sentence, in which Freud associates the father–son relationship with homosexual desire. In Genesis 9:24, symbolic resources are inversely related to such desire. Is it surprising that a "crisis" in either domain would destabilize the structure of the other?

This formula is discernible in Freud's developmental schema in which one's sense of social relationality becomes constituted across a series of differentiated stages of psychosexual organization, stages proceeding, presumably, from a narcissistic libidinal attachment to a socially relational one characterized by, in Santner's (1996, 53) words, a "passionate engagement" with "otherness." In this schema, homosexual desire bridges narcissism and the libidinal cathexis of otherness. Freud's idea that infants assume that everyone has the same genitals suggests, Santner (1996, 53–54, italics in original) comments parenthetically, that a "pre-oedipal boy's love for his mother is *homoerotic*," and doubly so, I might add, insofar the male infants' symbiotic union with heterosexually identified mothers would also internalize the mother's desire for men (including, presumably, the father). Freud imagines, then, that heterosexual desire does not replace homosexual desire; it is, in Freud's words, combined "with portions of the ego-instincts and, as 'attached' components, help to constitute the social instincts, thus contributing an erotic factor to friendship and comradeship, to *esprit de corps* and to the love of mankind in general" (quoted in Santner 1996, 53).

Employing the notion of "fixation" first offered in his earlier *Three Essays on the Theory of Sexuality*, Santner notes, Freud turns his attention to Schreber, suggesting that he, and paranoids generally, have never managed to move beyond a narcissistic homosexual desire. This failure to sublimate this desire into homosociality—a term Eve Sedgwick devises, as Santner acknowledges—left Schreber "exposed to the danger that some unusually intense wave of libido . . . may lead to a sexualization of their social instincts and so undo the sublimations which they had achieved in the course of their development" (Freud, 1953–1974, 62; quoted in Santner 1996, 54). I would add "racial" in conjunction to "social instincts" in that sentence. Certainly in the regression provoked by the gendered humiliation of the Civil War, many Southern white men (homo)sexualized their racial "instincts" into a ritualized form of castration: lynching.

Freud then suggests—I am following still Eric Santner—that the individual (living in a homophobic culture) defends himself against such desire by converting desire into persecution: "I do not *love* him—I *hate* him, because *he persecutes me.*'" The final clause, Santner (1996, 54) notes, is necessary because the "mechanism of symptom-formation in paranoia requires that internal perceptions—feelings—shall be replaced by external perceptions" (quoted in Santner 1996, 54). The negation of "feelings" does not conclude the matter, Santner (1996, 54) continues, as the "homophobic law" of this "disorder" requires their disavowal and projection onto the social world.

Significantly, Freud's theorization of paranoia suggests, Kaja Silverman (1988, 16) points out, that "vision" and "hearing" play "key" roles in the "relocation" of an "unwanted quality" from the "inside" to the "outside." More specifically, Silverman (1988, 16) continues, "visual" and "auditory hallucinations" perform a "critically important projective function," as the "projecting subject" protects himself against unwanted desire by repositioning what is unwanted at a visual and/or auditory distance by remaking what is unwanted internally the external "object" of the "scopic" and "invocatory drives." In so doing, (white) men protected themselves from specularity by converting it into scopophilia, providing them the sexual means by which they could derive pleasure from what they have, presumably, renounced (see Silverman 1988, 26).

Is this the basic white psychic structure of racism, a sexual and specular structure sketched in Noah's curse? Let us go in reverse: white men's obsession over what they imagine as black men's hypersexuality derives from the projection of their disavowed desire for the father, buried under their father's repudiation of that desire and, buried under that, the disavowal of Noah's own incestuous desire for the son. (Ham's desire is the desire of the Other and, as such, is "passive".[1]) In servitude and enslavement, white men claim complicity with the covenant—in which homoerotic desire is exchanged for genealogical generativity—while enjoying sadistic mastery over that desire they imagine as belonging to racialized others but which, in fact, is their own, displaced and deferred. The truth of heteronormative male desire is that it often has little to do with women; it is solipsistic and self-referential (see Silverman 1988, 143).

For Santner, Schreber's crisis was primarily a crisis of investiture. Schreber "discovered"—Santner's (1996, 124) verb may ascribe to Schreber a reflexive rational agency he did not exhibit—that the symbolic power and authority conferred upon him as judge and—as a "German man" Santner (1996, 124) adds—was performatively established. Sounding like Judith Butler, Santner (1996, 124) argues that Schreber's "symbolic function" was maintained by his capacity to produce a "regulated series" of "repeat performances," an "idiotic repetition compulsion" at the center of his symbolic function that Schreber experienced as "profoundly sexualizing," as, in fact, "jouissance."

This Lacanian reference Santner historicizes as follows: Schreber's experience of "jouissance" was not generically sexualized; it was for him feminizing and, in ethnic terms, "Jewifying" (Santner 1996, 125), suggesting, Santner offers, that at this historical conjuncture, "knowledge" of jouissance was, in Europe, ascribed to women and to Jews. This experience of abjection, Santner (1996, 125) continues, "signifies" a "cursed knowledge" of "jouissance," which, by means of "secondary revision," is decoded as "homosexuality," "femininity," or "Jewishness," what Schreber summarizes with the name *Luder*. Does, in this sense, racism become legible as a secondary revision of a cursed knowledge of jouissance, that knowledge mythically "discovered" by father and son inside Noah's tent, covered up and contained by the covenant?

Santner believes he is ready to revise those readings of the Schreber case made by Gilman, Geller, Boyarin, and Eilberg-Schwartz. Perhaps too eager to summarize, Santner (1996, 138) tells us that these scholars "all" argue that Freud's reading of the Schreber *Memoirs* was produced under pressure to disavow those "feminine" cultural elements that fin-de-siècle Central Europe associated with Jews. According to this reading, this association—doubly threatening to Freud—explains how he managed to miss, as Santner (1996, 138) nicely puts it, that "homophobia," not homosexuality, "produces paranoia." Likewise, it is the fear of and taboo against father–son incest, not incest itself, that inseminates racism.

Rather than reading Freud's missed opportunity as a consequence of his defensiveness regarding his own "femininity" and homosexuality (especially as it followed his relationships to Fleiss, Jung, and Ferenczi), Santner argues that the references to homosexuality and its "overcoming" (in his work on Schreber) should be contextualized as issues of originality and influence anxiety. Santner suggests that these issues—of paternity—were especially intense for Freud and his colleagues due to the fragile status of psychoanalysis as a science and institution. Moreover, Santner proposes that Freud's interest in the Schreber case was, finally, not a function of his identification with Schreber's presumed homosexuality, but, rather, with Schreber's struggle with a crisis of investiture, a "breakdown," as Santner (1996, 139) puts it, in the "*transfer* of symbolic capital" that would have enabled him to assume his position within the courts. This breakdown and Schreber's "hallucinatory" attempt to "repair" it, Santner (19996, 139) concludes, Freud "misread" as Schreber's homosexual desire for "paternal substitutes." Why these are not parallel, rather than mutually exclusive, readings is not obvious to me.

Does not the breakdown of identification and investiture in the symbolic order mean, on the psychological and specifically erotic level, a longing for the "big Other" with which one has failed to identify and which is now perceived as "somewhere else" or "someone else" and which now, in its separation, is longed for, rather than experienced as incorporated as "oneself"? And, read against Genesis 9:24, is not investiture always a "cursed" accomplishment, the consolation prize awarded to the sons who dare not see the "naked truth" of paternity, of gender, of race, of power?

The Panoptical Will to Knowledge

The notion of a God who demands our submission further reinforces the eroticization of dominance and control in men. (James B. Nelson 1996, 317)

Stereotyping through looks was basic to racism, a visually centered ideology. (George L. Mosse 1985, 134)

The prevailing Western concept of sexuality . . . already contains racism. (Kobena Mercer and Isaac Julien 1988, 106)

Santner reminds us that in the first volume of his *History of Sexuality* Foucault argued that the domain of functions, sensations, pleasures, and perversions known as human sexuality—the primary sphere of Schreber's "symptoms," Santner (1996, 84) reminds—was a complex consequence of an institutionalized "will to knowledge" the regimen of tests and examinations of which became, in the nineteenth century, obsessively focused on the body. The *History of Sexuality* was originally entitled "the will to knowledge" and contains, Santner notes, the only direct reference to Schreber in Foucault's *oeuvre*. Schreber signifies a general tendency of scientists, including behavioral and social scientists (fields just emerging in the late nineteenth century: see Baker 2001), to conceptualize a human life as a "case," one's life story as a "case history."

Foucault was suggesting that it was only in such a historical moment, characterized by "ever expanding regimes of expert knowledge or 'disciplines,'" among them "criminology, psychiatry, pedagogy, among others,"—that the Schreber case could first emerge and become the "classic" text that it became (Santner 1996, 84). Santner quotes Foucault to underscore the centrality of the examination in the formation by those disciplines interpellating the "individual" as a "describable, analyzable object" (1975, 190; quoted in Santner 1996, 84):

> The examination as the fixing, at once ritual and "scientific," of individual differences, as the pinning down of each individual in his own particularity (in contrast with the ceremony in which status, birth, privilege, function are manifested with all the spectacle of their marks) clearly indicates the appearance of a new modality of power in which each individual receives as his status his own individuality, and in which he is linked by his status to the features, the measurements, the gaps, the "marks" that characterize him and make him a "case." (1975, 192; quoted in Santner 1996, 84)

In our time, standardized examinations "mark"—even stereotype—the public school student in systems of social stratification that inevitably leave children behind.

For Foucault, Santner points out, the crucial model of the procedures of observation, examination, and registration that characterize the disciplines and therewith "constitute the individual as effect and object of power, as effect and object of knowledge" (1975, 192; Santner 1996, 85), was provided by Jeremy Bentham's *Panopticon* (see Pinar 2001, chapter 16). For Foucault, Santner points out, Bentham's design became not only an influential architectural design for an ideal prison in which the inmate would learn to internalize the agency of observation, it became the key metaphor for that technical rationality that emerged in the Enlightenment, "a figure of political technology that may and must be detached from any specific use" (1975, 205; quoted in Santner 1996, 85).

Foucault's analysis of panoptical discipline enables us to read Schreber's struggle with the "obscene father" (Santner 1996, 85) as a conflict between two conflicting legacies of the Enlightenment: the *liberties* and the *disciplines*. Santner (1996, 86) notes that Schreber's father's orthopedic treatments, pedagogical theories and practices, and public health and physical fitness programs amounted to a "caricature" of those systems of "micropower" that Foucault associates with the disciplines. Schreber's struggle with the "obscene, surplus father"—with God-the-Father—is read by Santner (1996, 84) as a struggle between the "law" and a "transgressive infra- or counter-law," which, Santner notes, Foucault characterizes as an "unassimilable residue" of "delinquency" (1975, 282; quoted in Santner 1996, 86). Santner quotes Foucault:

> [W]hereas the juridical systems define juridical subjects according to universal norms, the disciplines characterize, classify, specialize; they distribute along a scale, around a norm, hierarchize individuals in relation to one another and, if necessary, disqualify and invalidate. (1975, 222–223; quoted in Santner 1996, 85)

In late-nineteenth-century America, the case study individualizes and developmentalizes (in reverse sequence) widely held fantasies of racial regression;

in both instances, the individual disappears into codes of classification, distribution and hierarchization.

Foucault's most "Schreberian" insight, Santner (1996, 87) offers, is that disciplinary knowledge produces a new experience of the "intensified" body, one that recollects and travesties the sublime body. For Schreber as well as for Foucault, Santner (1996, 87, emphasis in original) notes, such an "intensification" of the body is "first" and "foremost" a "*sexualization*." That is, for Foucault contemporary hypersexuality is a consequence of, in Santner's (1996, 87) fine phrase, a "panoptical attentiveness" turned onto the "body" and its "sensations." For Foucault this attentiveness, Santner (1996, 87) points out, becomes institutionalized in medical, psychiatric, pedagogical, and other "professional" discourses, including in the appearance, in the late nineteenth century, of a "science" of sexuality. It is also institutionalized in eugenics (the "central tool" of which, Richard Dyer [1997, 104] suggests, was photography) and other racialized "sciences" (such as intelligence testing) structuring public education in the United States (see Selden 1999; Ravitch 2000).

Santner (1996, 87) discerns Foucault's description of this discursive production of sexuality in Schreber's struggle first with Flechsig's "tested soul" and then with the "rays of God," struggles that, Santner points out, "produced" intensifying "sexual excitation," ending in Schreber's "mutation" into a woman "completely saturated" by sexuality. Santner quotes Foucault:

> More than the taboos, this form of power demanded constant, attentive, and curious presences for its exercise; it presupposed proximities; it proceeded through examination and insistent observation . . . . It implied a physical proximity and an interplay of intense sensations . . . . The power which . . . took charge of sexuality set about contacting bodies, caressing them with its eyes, intensifying areas, electrifying surfaces, dramatizing troubled moments. It wrapped the sexual body in its embrace. There was undoubtedly an increase in effectiveness and an extension of the domain controlled; but also a sensualization of power and a gain of pleasure . . . . Power operated as a mechanism of attraction; it drew out those peculiarities over which it kept watch. Pleasure spread to the power that harried it; power anchored the pleasure it uncovered. (1976, 44–45; quoted in Santner 1996, 87)

Was racialized "servitude" the "anchor" that stabilized the "pleasure" uncovered inside Noah's tent?

Foucault's characterization of the "perpetual spirals of power and pleasure" (1976, 45; quoted in Santner 1996, 87), fashioned within the intimacies of medical examination, psychiatric investigations, pedagogical reports, and family controls, is recognizable in Schreber's experience of a God whose covenant with him is one of sexual stimulation, the "perpetual cultivation" of "jouissance" (Santner 1996, 87). Schreber's repeated requests that he be examined by scientists to confirm the spread of nerves of "feminine voluptuousness" throughout his body Santner understands as a contorted confirmation of Foucault's association of sexual agitation with intimate "scientific" examinations. Such sexual agitation was an example of what Foucault termed "biopower," in which knowledge/power transform human life, a fact for Foucault and, Santner adds, for Schreber, a defining feature of modernity. Santner (1996, 89) comments that Schreber's case fascinates us because it brings into such "sharp relief" an era of "crisis" in the historical "tension" among "forms" and "systems" of "power" and "authority."

These forms and systems become vivid in Schreber's gendered spiritual visions. In order to see himself as a woman, Gilman points out, Schreber had used his imagination: "The picture of the female buttocks on my body . . . has become such a habit that I do it almost automatically whenever I bend down" (1968 [1903], 181; quoted in Gilman 1993, 159). "By becoming a woman," Gilman (1993, 160) suggests, "Schreber undertook what was expected of a Jew—that the Jew become someone else, with a different body, in order to become new." In these moments of "seeing-as," Sass (1994, 123) points out, Schreber's extent of feminine identification "waxes and wanes" according to the extent to which he becomes the object of God's gaze.

To illustrate his claim that Schreber was self-aware of women's objectification through the (straight) male gaze, Louis Sass (see 1994, 123) juxtaposes a passage from the *Memoirs* with John Berger's famous formula (also recalled by Howard Eilberg-Schwartz [see 1994, 95]). Schreber reported that while "male voluptuousness" is "stimulated" by the "sight" of "female nudes," "female voluptuousness" responds to the sight of both (1968 [1903], 142; quoted in Sass 1994, 123). Sass concludes that Schreber's ambivalence about becoming a woman cannot be grasped apart from the epistemological relation between visuality and identity. This is an epistemological relation that is simultaneously racialized and gendered, as Gilman makes clear.

Sass focuses on the gender of the relation, suggesting that becoming a woman, for Schreber, means "ceding one's epistemological centrality and becoming a mere object defined by the other's sovereign awareness" (Sass 1994, 123). But "epistemological centrality" and "sovereign awareness" may well be solipsistic male fantasies best "ceded" as soon and as completely as possible. There are the privileges of Shem and Japheth, the "circumcised" sons cursed by accepting the covenant between father and son.

In another passage, Sass (1994, 26) points out, Schreber describes a process of "picturing" as volitional and self-conscious: "To picture (in the sense of the soul-language) is the conscious use of the human imagination for the purpose of producing pictures (predominantly pictures of recollections) in one's head, which can then be looked at by rays" (1968 [1903], 180–181; quoted in Sass 1994, 26). That is, "pictures" in one's mind enable Schreber—and God—to transform material reality.

> I can also "picture" myself in a different place . . . standing in front of a mirror in the adjoining room in female attire, when I am lying in bed at night I can give myself and the rays the impression that my body has female breasts and a female sexed organ. (1968 [1903], 181; quoted in Sass 1994, 27)

Is "visuality" here a masculinized retreat into the world from what is felt to be a feminine "inner" reality?

DEFLECTING THE DESIRE OF GOD

The delusional system of the paranoid is equivalent to the organized system of theology. (Sander L. Gilman 1993, 146)

Do we have to have only one point of exit from the kingdom of the phallus? I think, on the contrary: the more, the merrier. (Rosi Braidotti 1994a, 53)

To desire the Jesus who is not white, the Jesus who is black, is to place desire in the service of overcoming domination. (Robin Hawley Gorsline 1996, 135)

Freud's *Moses and Monotheism*, Eilberg-Schwartz reminds us, is devalued by many as a product of his old age, composed while preoccupied with his deteriorating health, the fragility of the psychoanalytic movement in Vienna, and the vulnerability of Jews in Europe. Because the main argument of *Moses and Monotheism* tends not to be taken seriously, Eilberg-Schwartz notes, scholarly attention (he is thinking especially of the work of Jay Geller; I think of Sander Gilman and Daniel Boyarin) has recently focused instead on what this work says about Freud's own Jewishness. Eilberg-Schwartz focuses on the third essay in the book, which contains, he suggests, Freud's main claims about monotheism.

In the context of reviewing his historical argument about Moses' Egyptian origin and end, Freud elucidates his understanding of psychological latency and repression, processes presumably parallel to the development of monotheism. Reversing the nineteenth-century Herbartian claim that ontogeny recapitulates phylogeny, Freud argues that the primal crime against Moses and the religion of Moses is repressed but returns in the history of the Jews, just as childhood memories come back to haunt an adult. Freud recounts the tale of primal parricide he depicted in *Totem and Taboo* wherein the father is hidden in the symbol of an animal. Those religious practices of worshipping and ritualistically eating animals express, Freud suggests, a reverence for and fear of the father. In monotheism the presence of the absent father in God-the-Father is made explicit:

> [T[he first step from totemism was the humanizing of the being who was worshipped. In the place of animals, human gods appear, whose derivation from the totem is not concealed . . . . The male deities appear first as sons beside the great mothers and only later clearly assume the features of father-figures . . . . The next step [the development of monotheism], however, leads us to the theme which we are here concerned—the return of a single father-god of unlimited dominion. (1939, 106; quoted in Eilberg-Schwartz 1994, 31–32)

Is "unlimited dominion" the spiritual compensation for men's sexual emasculation, that is, castration?

For Freud, Eilberg-Schwartz reminds us, monotheism was the first religion to recognize the father behind the experience of God: "The re-establishment of the primal father in his historic rights was a great step forward but it could not be the end" (1939, 86; quoted in Eilberg-Schwartz 1994, 32). Freud then contrasts the religion of Moses with the religion of Jesus. If the Mosaic religion represents the cult of the Father, Christianity is the cult of the son. Did lynching recapitulate the deferred and displaced intra-psychic violence of "self-made" men—fathers murdered, mothers denied—who must act out repressed desire for the father and disavowed identification with the mother during the late-nineteenth-century racialized "crisis" of masculinity?

Two thousand years ago the Father may have forsaken his son on the cross, but in the cult that the crucifixion symbolizes the son is resurrected. In this sense, Eilberg-Schwartz points out, Christianity represents a regression in Freud's language, for it undermined the centrality of the father God, ended strict monotheism, and replaced it with the notion of the Trinity, curiously the same number present in Freud's reconstruction of the—sodomitical, as Edelman observed—primal scene. In another sense, the fantasy of Christ's resurrection is a return of the primal father. Through the crucifixion of the son, Christians were, presumably, able to atone for the original sin, the murder of the father, a variation on the Noah story wherein the enslavement/crucifixion of his son's sons represents atonement for the unmanning of the father. Freud is focused not on white racism, but on anti-Semitism: although Christians acknowledged and atoned for their original sin, namely, killing the father god, Jews never would (Freud 1939; Eilberg-Schwartz 1994).

Another of Freud's claims that concerns Eilberg-Schwartz is the argument that a conceptual, cultural, and moral revolution occurred when Moses introduced the Jews to an abstract conception of God. The prohibition against making divine images, (Freud suggested and Eilberg-Schwartz reminds), presumably permitted Jews to transcend the senses, given that they could no longer envision God in human or sensate form. This triumph of spirituality over the senses amounted to (racialized) renunciation of the instincts, for Freud a sign of maturation and progress. Once again phylogeny recapitulates ontogeny: Freud underlined the parallel between the transformation of a people and the renunciation of instinctual gratifications a boy must accomplish on his way to manhood and maturity:

> The religion which began with the prohibition against making an image of God develops more and more in the course of the centuries into a religion of instinctual renunciations. It is not that it would demand sexual *abstinence*; it is content with a marked restriction of sexual freedom. God, however, becomes entirely removed from sexuality and elevated into the ideal of ethical perfection. But ethics is a limitation of instinct. (1939, 148; quoted in Eilberg-Schwartz 1994, 33)

For Eilberg-Schwartz, these are Freud's most interesting observations about monotheism, namely his associations among the fatherhood of God, the prohibition of images, (homo)sexual renunciation, and the (presumed) triumph of the spirit over the senses.

On this last point, Eilberg-Schwartz is skeptical: where Freud linked the prohibition on images of God with a triumph of spirituality over the senses, Eilberg-Schwartz understands it as a means of deflecting homoerotic complications raised by the fact of God's—the Father's—(male) body. It is here that Eilberg-Schwartz (1994, 33) tells us that his own account of monotheism and the centrality of the homoerotic dilemma is "Freudian through and through." It is a Freudian interpretation of monotheism, he continues, that Freud himself did not formulate (see Eilberg-Schwartz 1994, 33). Does the son—through his self-abjection in worship—testify to the sexual desire the Other (Father) denies?

To formulate his reading of monotheism, Eilberg-Schwartz reviews Freud's theory of the Oedipus complex. He notes that Freud postulated an original bisexual desire in the newborn that only later becomes directed primarily toward the "opposite" sex. The so-called positive oedipal resolution is central to *Totem and Taboo* and *Moses and Monotheism*. In this "positive" outcome the male child desires (only) his mother; due to that desire he worries his father will punish—castrate—him, an event Freud believed occurred repeatedly in history, (psychological) events of which circumcision may be understood to be a sublimated residue. Terrified he will lose his penis, the son suppresses his desire for his mother and cooperates with his father, a castration of another sort one might add. Never mind, Freud regarded this outcome as the successful resolution of the Oedipus complex (Eilberg-Schwartz 1994).

Freud knew there was another object of desire for the son, one which, if not repressed, could result in a rather different outcome, what he called an "inverted" or "passive" or "negative" Oedipus complex. In this version, the boy remains identified with his mother, and, in this "castrated" position, desires to replace his mother and become the object of his father's affections. (Noah would seem to be the inversion of this fantasy, in that the son "castrates" the father, although we cannot rule out the possibility that it was Noah who was the sexual aggressor.) This desire to become a woman and take the (heterosexual) female position can have a "successful" outcome (as it did not for Schreber), provided the son renounces this desire for his father and identifies with him instead, copying his father's presumed desire for an "opposite-sexed" love object. If this desire for the father is not renounced, if identification is not achieved, Freud believed that paranoia or other disorders, including homosexuality, may follow (Eilberg-Schwartz 1994).

Although the relationship between the worship of God and sexual longings for the father are not taken up in *Totem and Taboo* or in *Moses and Monotheism*, Freud analyzed them on other occasions, most notably in his case studies. Eilberg-Schwartz cites Freud's study of Schreber as an example of a case of paranoia that involved homoerotic feelings toward a father God. For Freud, Eilberg-Schwartz suggests, Schreber's case illustrates what goes wrong if a boy does not successfully resolve his oedipal desires for his father.

CHAPTER 7

The Curse of the Covenant

THE SYMBOLIC SUBSTITUTE OF CASTRATION

How far can a notion like castration be stretched or resignified before it loses its structural value and epistemological effectivity? (Teresa de Lauretis 1994, 308)

And instead of symbolizing, like the phallus of Dionysus, the generative powers of nature, Christ's sexual organ—pruned by circumcision in sign of corrupted nature's correction—is offered to immolation. (Leo Steinberg 1996 [1983], 47)

The theory of interpellation appears to stage a social scene in which a subject is hailed, the subject turns around, and the subject then accepts the terms by which he or she is hailed. (Judith Butler 1997, 106)

The internalization of interpellation is none other than such an (irrational and senseless) act of patching over, of stitching together this chasm. (Rey Chow 2002, 109)

While early anthropologists labored to describe the tribal rites they observed, the first generation of psychoanalysts, most notably, Sigmund Freud, focused on their symbolic significance, specifically the meaning of the ritual of circumcision. Gollaher notes that as a Jew, Freud himself had been circumcised as an infant, although he did not participate in Judaic rituals as an adult. What intrigued him, Gollaher notes, were connections between cutting the penis—as an anatomist he could not consider the foreskin a separate structure—and his evolving theory of sexuality. Especially pertinent here was the relationship—one of "deferred action"—between childhood trauma and later neurosis (Gollaher 2000; Lukacher 1986).

Freud knew of Maimonides' opinion (see chapter 2, "A Very Very Hard Thing") that circumcision was meant to be traumatic, as its practical purpose was to inhibit male sexuality. Freud agreed that circumcision did indeed accomplish this, but, unlike Maimonides, he did not endorse this "accomplishment." By the mid-1890s, Freud had become an advocate of sexual expression, convinced that physical or emotional repression of sexual arousal and release engendered anxiety and neurosis (Gollaher 2000). "It is positively a matter of public interest," he declared (anticipating his future ex-colleague Wilhelm Reich), "that *men should enter upon sexual relation with full potency*" (quoted in Gollaher 2000, 67).

Given his view that circumcision did inhibit sexuality, Freud was, predictably, critical of it. His criticism—Gollaher (2000, 67) judges it "extravagant"—seems to me right, if framed within his theory of the "positive" oedipal complex:

> Circumcision is the symbolical substitute of castration, a punishment which the primeval father dealt his sons long ago out of the fullness of his power; and whosoever accepted this symbol showed by doing so that he was ready to submit to the father's will, although it was at the cost of a painful sacrifice. (Quoted in Gollaher 2000, 67)

With an identificatory oedipal politics, castration represents submission to the father's will, his participation in patriarchy, concretely enacted by Shem and Japheth's cover-up of their father's "nakedness." But given the unstable binary identification/desire, it could also suggest Ham's submission to his father's desire.

In either case, the father seems "jealous" and "cruel," or so Freud imagined. Gollaher notes that Freud's theory of oedipal origins became conflated with his speculation concerning the prehistoric past. In both *Moses and Monotheism* and his *New Introductory Lectures on Psychoanalysis*, Freud (1939, 192; 1933, 120–121) claimed that "in the early days of the human family, castration was performed on the growing boy by the jealous and cruel father, and that circumcision, which is so frequently an element in puberty rites, is an easily recognizable trace of it" (quoted in Gollaher 2000, 67). The mark of the father on the son becomes converted into the homosocial "covenant" of fraternal fascism.

As Gollaher observes, a number of American psychoanalytic theorists, especially in the United States, worked to elaborate a universal theory of circumcision. He notes that Theodore Reik constructed circumcision as a kind of anticipatory punishment in which older members of the family or community reprimanded young men for their "secret sexual desires" (Gollaher 2000, 67). Psychoanalytically influenced anthropologist John Wesley Mayhew Whiting suggested that the trauma of circumcision functioned to break the incestuous bond between mother and son, providing a means for the son to enter the adult male world without a parricidal revolt against the father. Reiterating the common complaint that they are reductionistic, Gollaher asserts that psychoanalytic interpretations largely ignored circumcision's social and ritual significance, as if ritual and symbolism were only incidental to a "deeper psychological truth" (Gollaher 2000, 67).

Gollaher's hostility to psychoanalysis—he finds its arguments "improbable" (2000, 68)—is evident in the tone (one of contained incredulity) of his gloss of Herman Nunberg's (1949) *Problems of Bisexuality as Reflected in Circumcision*, in which Nunberg asserted (quoted in Gollaher 2000, 68) that the "study of puberty rites of primitives proved that circumcision represents symbolic castration, its underlying motive being prevention of incest." (Not disagreeing, I suggest circumcision represents a sublimated sign of incestuous desire between father and son, a ritualized trace of the "negative" oedipal complex, converting the sexual into the sacred.) Drawing on his own work with patients, Nunberg concluded that circumcised boys tended to blame what they felt was a castrating experience upon their mothers and, as a residue of the trauma, continued to experience hostility and guilt (Gollaher 2000). This is a curious sequence, in that circumcision is a ceremony conducted on boys by men in the name of the Father, sacred or secular. True, mothers, in effect, "agree" to the ceremony, and it is possible that sons might feel some sense of betrayal if they imagine the mother as "phallic" and capable of protecting them (but declining to do so), but the ascription of responsibility to her seems strange and "overdetermined."

Nunberg also understood circumcision as an expression of deep-seated anxiety about gender. After Freud, he appreciated that cutting the foreskin was cutting the penis, and this mutilation of the penis permitted the male to become female-like in his genitalia, a point Gollaher noted earlier in his review of tribal circumcision. Others took exactly the opposite position, arguing that circumcision stemmed from "the wish to create in males a permanent erection of the penis to ensure . . . fertile sexuality and thence the continuity of the group" (quoted in Gollaher 2000, 68). Of course, apparently opposite explanations can both be true. If we appreciate that castration anxiety engenders hypermasculine defensiveness and compensation, the fantasy of the cut penis as both female-like and constantly erect becomes perfectly plausible. Although Gollaher declines to take psychoanalytic theory seriously, he does, patronizingly to be sure, credit it with inspiring Freud's successors to investigate the "psychological impact" of the procedure (2000, 66).

One investigation that Gollaher cites took place in Turkey, where researchers studied twelve boys in a mental hospital. They questioned each child's mother about her son's environment, emotional, social, and intellectual development, then administered the Goodenough draw-a-man test, Rorschach blots, and CT scans to the children before and after circumcision. They found that after circumcision these boys seemed to regress, drawing themselves as smaller and younger than they had in the images they had drawn earlier. "The operation is experienced by the child as an aggressive attack, with deadening implications," the researchers concluded.

> The results obtained for the different psychological tests indicate that circumcision is perceived by the child as aggressive attack on his body, which damaged, mutilated and in some cases totally destroyed him. The feeling of an "I am now castrated" seems to prevail in the psychic world of the child. (Quoted in Gollaher 2000, 68)

These findings appear to corroborate Anna Freud's views on childhood circumcision, Gollaher concedes. A founder of child psychoanalysis (see Young-Bruehl 1996; Britzman 2003), Anna Freud believed that boys experience little difference between circumcision and castration; consequently, circumcision undermines boys' psychological development (Gollaher 2000). The "helplessness," "deficiency," and "physical shrinking" of "self-image" observed in Turkey, Gollaher (2000, 68) observes, "seemed to prove her right."

Gollaher (2000, 68) judges as the "most ambitious" effort to provide a psychological explanation for circumcision the 1950s work of Bruno Bettelheim. "Whatever the origin and meaning of circumcision may be, it must originate in deep human needs," Bettleheim wrote, "since it seems to have sprung up independently among many peoples, although in different forms" (quoted in Gollaher 2000, 68). Why, he asked, would so many diverse peoples take up such a radical and risky operation? It was "a strange mutilation," all the stranger for its being "found among the most primitive and the most civilized people." Circumcision, he concluded, "must reflect profound needs" (quoted in Gollaher 2000, 69). What could those "profound needs" be?

Bettelheim dismissed Freud's speculation regarding a primal castration. Rather than anticipatory punishment for desire yet to be experienced, Bettelheim argued that circumcision reflected a deep-seated ambivalence about being confined to a single sex: "The desire to possess also the characteristics of the other sex is a necessary consequence of the sex differences." The satisfaction of this desire would mean, however, the loss of one's own genitals, hence the inexorable nature of castration anxiety in both sexes." To make this fear manageable, circumcision and other rites of initiation were devised, a curious and certainly culturally circumscribed hypothesis given the terrorization of boys that obviously occurred among some tribes. Bettelheim insisted that circumcision was not about castration, but fertility. Against Freud, Bettelheim argued that "circumcision developed as a result both of man's desire to participate in the female power of procreation, and of woman's desire, if not to rob the male of the penis, at least to make him bleed from his genitals as women do" (quoted passages in Gollaher 2000, 69). It seems to me that he is arguing for a generalized female-envy, not fertility-envy only.

Bettelheim's conclusions were based on his reading of scholarly literature and upon his work with mentally disturbed boys at the Sonia Shankman Orthogenic School of the University of Chicago. Blood fascinated these boys, Bettelheim noticed, and several of them imagined menstrual blood as magical. Circumcision and the bleeding it produced appealed to boys because it seemed to promise them the magical powers girls enjoyed, among them, the capacity to procreate. In what Gollaher (2000, 69) characterizes as a "flight of fancy," Bettelheim likened their experience to Australian aboriginal circumcision rituals, in which boys are separated from their mothers, held captive by older men, and finally, through the initiation ritual, "reborn as men." What the aboriginal ritual signified, Bettelheim speculated, was that circumcision, particularly its blood and pain, provided men the illusion that they possessed the procreative power of women. Although they were in fact unable to produce babies, "men, through the transforming circumcision ritual, demonstrate the power to produce men" (Gollaher 2000, 70). Beginning with the opening verses of Genesis—which "give powerful expression to the fantasy of creating something out of nothing" (Silverman 2000, 20)—(especially white) men have endowed their capacity to (pro)create (patriarchal) culture with the prestige of biological reproduction (see, also, Halperin 1990, 144).

Bettelheim's realization does not, it seems to me, necessarily contradict Freud's emphasis upon castration, if we understand castration anxiety as the sneaking suspicion that men are already castrated, already not-men, already "women," which preoedipally they are, symbiotically merged with the maternal body from which they were physically separated at birth. The circumcision ceremony, in this respect, resembles a reminder that men are always already castrated, that they envy women of whom they are negated, abjected copies. Circumcision reminds men whom they are no longer, cannot be, but whom they will try to possess: women. "The core object of self-attack," Joseph Smith (2004, 351) asserts, "pertains to the wish-fear of reunion with the early mother."

Is, then, castration this loss of the symbiotic identification with the mother, the loss of primordial wholeness? Is men's castration of each other a secondary wound, reiterating, if deferred and displaced, the primal wound of birth and separation? Men know they are not "men," like whiteness an "ideal" by definition incapable of mimesis, fated to fail (DiPiero 2002). In prison, through rape, men can force men into "becoming-woman" (Pinar 2001, chapters 16 and 17). Homophobia itself, Leo Bersani (1995, 77) speculates, "may be the vicious expression of a more or less hidden fantasy of males participating, principally through anal sex, in what is presumed to be the terrifying phenomenon of female sexuality."

"If there is a certain logic in such reasoning," Gollaher (2000, 70) allows, once again revealing his patronizing hostility to psychoanalytic theory, "ultimately Bettelheim's theories collapse into a pile of conflicting conjectures." As "evidence" for this conclusion, Gollaher cites Bettelheim's "confession," made at the end of his study: "there is much evidence that it is imposed or desired by women; but there is also much reason to believe that it is desired by men because it (a) it makes them more male by freeing the glans; (b) it provides them with a sign of sexual maturity and with potent blood from the genitals; and (c) it adds to their power by giving them symbolically the capabilities of women" (quoted in Gollaher 2000, 70). Quite aside from the epistemologically complex issue of "evidence," these are hardly self-contradictory conclusions. Men often want it both ways: in fact, they want it every which way they can.

Not only does Bettelheim's work "collapse into a pile of conflicting conjectures," it is only a partial pile at that, as Gollaher (2000, 70) alleges that Bettelheim "missed what scholars in the field" had long ago discovered, namely, that among certain African groups—in the Dogon, Bambara, the Lodi of Mali—the cutting of male and female genitalia reflects their belief in the fundamental duality of human beings. The fact that belief does not always equal explanation seems to escape Gollaher at this point. Curiously, his observation seems congruent with Bettelheim's acknowledgement that men want to be both men and women, as he reports that in some tribes, a newborn is said to possess twin souls of both sexes. In girls, the masculine soul presumably inhabits her clitoris, for years in the West imagined as a tiny penis. It is removed so she may become feminine. In boys, the female soul lives in the foreskin; circumcision frees boys to become "men." Gollaher (see 2000, 70) quotes social psychologist Pierre Erny, who reports, in his study of African children, that "after circumcision it is the man's duty to go after his lost femininity and find it again in his wife. And the woman who was freed from her masculinity at the time of excision finds it again in the person of her husband."

Gollaher (see 2000, 70) also quotes ethnographer Dominique Zahan's study of certain northwest African tribes: "In the spiritual realm the function of circumcision is still more nuanced. By circumcising man the blacksmith (who customarily performs the operation) takes away the 'femininity' from his spirit, that is, the cloudiness in his understanding, the *wanzo*." In the mythology of the tribes Zahan studied, *wanzo* represents ignorance and pollution, preventing a man from knowing himself and from knowing God. Self-knowledge and religious understanding require the excision of *wanzo*. Women would only marry men who are without *wanzo*. But because *wanzo* is a man's femininity, losing it deprives him of an essential element of himself. Marriage thereby restores to a man what he lost in circumcision (see Gollaher 2000, 70). By this logic, if boys were not circumcised, two men could compliment each other. The resonance with Adam's rib is redolent.

The imaginary status of women in men's minds is obvious here: femininity is what men already have until they shed it in order to be "men." But they must have it back, if in split-off and culturally regulated fashion: heterosexual marriage. That women, as concrete, singular subjectivities who might choose rather different trajectories of coupling and union (namely with themselves) is unfathomable to many men who project their own gendered and sexual duality onto the world they must then control, so that the world coincides with their fantasy of it. Bettelheim understood that men want to be "all they can be," and so they castrate each other in order to bleed, to signify women. But in so doing they also expel the feminine nature the presence of which informs their manhood. What is left is, magically, paradoxically, not a castrated penis or a large clitoris, but the "phallus," a penis continually erect, at least in men's minds, a badge of manhood and power used to subject not only women, but boys and other men as well.

This is no "universal theory," Gollaher (2000, 71) appreciates. "No theory fits the myriad facts," he writes on the same page, but this statement seems inspired by the dream of a mimetic epistemology (Trueit 2005). No theory can ever fit the "facts," as all theorization is embedded in the theorist's culture and historical moment, in his or her subjective situatedness, fashioned by his or her intellectual project and political passions. But that fact hardly relegates theory to "piles of conjecture." It means that theoretical acts are cultural and political acts, and although the truth of their assertions is socially, historically, and culturally contingent, it can carry the force of "truth" in an ongoing human conversation about the nature of the world we inhabit and construct (Oakeshott 1959). The "force" of truth is political and not, as Jürgen Habermas knew, without distortion, but the phrase does communicate the sense of engaged thinking that animates the conception of curriculum as complicated conversation (see Pinar 2004a).

In the Name of the Father

[T]he story of Noah suggests that averting the eyes from the father's nakedness is paradigmatic. (Howard Eilberg-Schwartz 1994, 119)

[S]ex with the father impinges on his authority. (Regina M. Schwartz 1997, 108)

The white subject is magnetically aligned with other races . . . only under conditions of disembodiment. (Russ Castronovo 2001, 167)

Castration links . . . gender and race. (Gary Taylor 2002, 146)

"Our present dominant fiction," Kaja Silverman (1992, 34) argues, "is above all else the representational system through which the subject is accommodated to the Name-of-the-Father." Its most central signifier of unity is the patriarchal family, she suggests; its primary signifier of privilege is the phallus. "Male" and "female" constitute our "dominant fiction's" most basic binary opposition (Silverman 1992, 35). She links various other ideological elements, such as "town" and "nation"[1] or the antithesis of power and the people, in a metaphoric relation to this basic binary of gender. It is from that oppositional relation, Silverman argues, that these other signifiers derive their conceptual and affective value, a point differently made by Eve Kosofsky Sedgwick (1990).

Lacan equates culture with the Name-of-the-Father. "In all strictness, the symbolic father is to be conceived as 'transcendent,' as an irreducible given of the signifier," Lacan asserts in his seminar of March–April, 1957. Silverman (see 1992, 7) quotes Lacan: "It is in *the name of the father* that we must recognize the support of the symbolic function which, from the dawn of history, has identified his person with the figure of the law." Is "naming" a sublimated derivative of desire, claiming an identificatory position in a now racialized intersubjectivity, codifying conduct so that desire is constrained, rerouted, mandated?

Juliet Mitchell agrees with Lévi-Strauss and Lacan on this point, Silverman notes, writing that "the systematic exchange of women is definitional of human society" (quoted in Silverman 1992, 7). Mitchell asserts that the Name-of-the-Father is synonymous with culture. Certainly it is synonymous with European culture, but not only, as David Gilmore's (1990) summary of anthropological research suggests. In the West, older men write on the bodies of young men, inscribing the Law-of-the-Father through circumcision.

They write on the minds of young men through schooling. In several tribal societies, the Name-of-the-Father is inscribed through even more explicit and violent forms of ritualized homosexual assault.

If we understand circumcision as marking "manhood"—an identificatory claim by means of which desire *for* the father is converted to desire to *become* the father—then we realize that compulsory *heterosexuality* is the founding moment, the primal scene, of misogyny, inextricably linked, as it tends to be, with fraternal fascism. It is compulsory heterosexuality, Trevor Hope (1994, 194) argues, that has substituted self-surveillance for sex, self-policing for self-exploration and, I would add, a fetishized commodity capitalism for the democratization of social relations.[2] Hope (1994, 194) insists that a close reading of *Totem and Taboo* discloses that the father becomes "installed" as the "Other," by which Hope means the law, "precisely" in its absence of subjectivity. The social body becomes, then, saturated with a "paranoid economy" structured by the "policing" gaze, not incoincidentally just as male homosexuality becomes "installed" or "retro-jected" as "irretrievably lost" by virtue of a "founding disavowal" (Hope 1994, 194). The (male) policeman's gaze—Althusser's moment of appellation—when directed at other men, performs the disavowal of desire through identification.

In the primal scene that is Genesis 9:24, Ham's visual encounter with his father's naked body—even if that is "all" that happened inside Noah's tent—is not policed. When the other two sons—Shem and Japheth—appear, they resolutely refuse to look at the father. In so doing, their incestuous desire is disavowed; it operates as self-renunciation, as self-policing and surveillance. This is the founding moment of patriarchy, that consolation prize awarded to those sons who have renounced their desire for the father in return for identification and power. The sons decline to see him (and themselves) as embodied sexual others, demanding to see it in others.

After the covenant, the sons gaze aggressively at the mother and her substitutes, a gaze now overdetermined by compounding their desire for her with their denied desire for the father, intensified by their rage that she is not he. "Why can't a woman be more like a man," Professor Higgins asks. Is misogyny, then, also repressed or frustrated homosexual desire? Hope (1994, 194) seems to imply as much, noting that the brothers—"fraternal citizen-subjects," as he aptly names them—demonstrate their loyalty to the "paternal corpse" by means of their "deferred obedience." Such an "encrypting" of the relation between sons and fathers in the "homophobic" and "misogynist ritual" of the "totemic meal" enables them to "in-corporate" the father "precisely" through the evasion of "*having him*" (Hope 1994, 194, emphasis in the original). In this formulation, homosociality is the repression of incestuous homosexual desire, and its melancholic residue of that night inside Noah's tent leaves the descendents of the sublimated sons—Shem and Japheth—in solidarity but angry, ready to murder each other and violate the women they keep "in the tent," in enforced domesticity, in compulsory femininity.

Misogyny cannot be ascribed to homosexual desire, then, but to its denial. Misogyny follows from those masculine identificatory politics that intensify a nearly universal male disavowal of the feminine. In the conclusion to his cross-cultural study of misogyny, David Gilmore (2001, 230) advises men to become "more comfortable with their ambiguous sexuality, their subterranean dependency needs for women's nurturing, their 'corrupt' feminine side, and their 'poisonous' bisexual self." (The quoted adjectives refer to characterizations of the feminine made by various tribal peoples studied by the anthropologists Gilmore discusses.) Gilmore calls for men to accept these "supposed weaknesses as normal and not tantamount to emasculation" (2001, 230). Stop denying "lack," as men narcissistically, defensively, experience the "feminine," that "rib" displaced from their psychic bodies onto the sexualized "other."

Only through such self-restructuring, Gilmore (2001, 230) suggests, can men learn to love women "unambiguously." It is only by reconfiguring one's self, by appreciating the "womanish" (2001, 230) in themselves, can men accept what they have projected onto women: the need for love, dependency, and the longing for comfort. Gilmore (2001, 230) asserts that "only self-knowledge can free men from the fear of women, and self-knowledge in this case means the acceptance of the divided self within and an imperfect universe without." (Such acceptance is easier for men to make given their political dominance nearly worldwide.) Then, in a discursive turn that links, through visual imagery, the gendered and the racialized, Gilmore (2001, 230) concludes:

> The key lies in forging a primarily alliance with the self and an acceptance of the incongruities of human existence. The demon within thus neutralized by self-forgiveness, one can then go on to form an alliance with the Other against the darkness . . . . Men and women thus united, misogyny will wither away of its own accord.

Is he suggesting that Adam reinsert his "rib"? Would the Garden then be free of (black) serpents?

Hope (1994, 194) agrees that "the" "social contract" is a "pact" of "fraternal concord" that "destitutes" the feminine. But Hope's attention is turned not to the destitution of the feminine, but to how this homosocial–patriarchal "pact" produces a profound melancholy in the heteronormative male subject, due to the evasions and deferrals accompanying the repression of homosexual desire. Homosexual repression—individually performed and socially structured as homosociality—"seduces" men when they "fall" for its "ruse" of "pathos," substituting melancholy for the excitement of sexual adventure. Patriarchy, Hope (1994, 194) concludes, amounts to a "representational - *pere-version*" socially organized as the "brotherly regime" of "modernity" (Hope 1994, 194). Modernity is the regime of racialization.

According to Hope's (1994, 194) analysis, male homosexuality is not a logic that founds modernist misogyny, but an "abyssal ground" that "confounds" men and "launches" them into their "phobic existence." The feminist analytic that insists on locating it at the origin is, Hope argues, complicit in its repression, a political symbolic that relegates homosexual desire to history while forbidding it in the present. David Eng (2001, 129) also points out that the "normative (re)construction" of the "primal scene" requires the "reworking of a homosexual presence into a homosexual absence" and the "strict segregation of identification and desire." Echoing Lee Edelman, Eng (2001, 129) characterizes this phenomenon as a kind of "reverse fetishism" or "reverse hallucination," in simple terms, "not seeing what is there to see."

In Hope's association of melancholy (that is, unresolved mourning) and sexualized phobia is an echo of Freud's original formulation—based on his reading of Schreber's *Memoirs*—of paranoia as the repression of homosexual desire. Although the idea is too general, too reductive, it does speak specifically of the curse of Ham. Noah does not displace his desire onto God who then seeks him: he becomes his repudiation through the performance of rage, an emotional act enabling him to displace his disavowed desire onto "others" now cursed to perpetual servitude, in his possession. Castrated by the covenant, Christian sons long for God-the-Father but the compensatory fantasy fails, and so they search for the enfleshed phallus, phobically finding it, they imagine, in Africa.

THE FUNCTION OF THE PHALLUS

The African has become not only the Other who is everyone else except me, but rather the key which, in its abnormal differences, specifies the identity of the Same. (V. Y. Mudimbe 1988, 12)

[T]he Infant's penis is not merely revealed, but pointed to, garlanded, celebrated. Soon after c.1500, it may even be touched and manipulated. At last (notably in the 1530s . . . .), Christ's male member asserts itself in erection—patently so in the babe, undercover in Christ dead or risen. (Leo Steinberg 1996 [1983], 225)

What caused Europeans to focus on skin color, particular facial features, or hair when trying to determine whether a specific cultural group was principally like or unlike them? (Thomas DiPiero 2002, 6)

Although the dark skin associated with Africa was known in Europe at least as far back as the sixth century B.C.E. (as indicated in the often-cited passage from Jeremiah 13:23: "Can the Ethiopian change his skin or the leopard his spots?"), DiPiero (2002, 63) reports that, in general, the ancients did not employ "color" and other apparent morphological differences to demarcate moral and intellectual superiority. He notes that the ancient Greeks portrayed the Ethiopians and other dark-skinned peoples as physically and culturally unusual in Greek poetic literature as early as Homer and Hesiod, and in other historical documents. In his *Histories*, Herodotus depicts the Ethiopians variously as "the tallest and best-looking people in the world" and the "longest-lived often attaining the age of one hundred and twenty" (quoted in DiPiero 2002, 63).

I would point out that Herodotus is focusing on the black body, not black subjectivity, and, moreover, its size; in the second adjective, he expresses his erotic response to the body on which his eyes have fastened. Richard Dyer (1997, 22) observes that, over time, "non-whites" became "seen as degenerate," a notion that "goes back at least to Johan Boemus, who in 1521 proposed that all humans were descended from Ham, Shem, and Japheth, the sons of Noah, but those who descended from Ham degenerated into blackness." Such a view rationalized the European rape of Africa.

The most active period of European colonization lasted less than a century. Other events associated with colonialism, which involved the greater part of the African continent, occurred between the nineteenth and mid-twentieth centuries. From the point of view of African history, the colonial experience represents but a brief moment. Still, this moment remains charged today, since it precipitated unspeakable suffering, incalculable destruction, rationalized by the production of new discourses on African traditions and cultures. One would have thought that the events of the past four hundred years would have erased two contradictory European myths, but they remain, in some attenuated form, today (Mudimbe 1988). The first might be termed the "Hobbesian picture of pre-European Africa, in which there was no account of Time; no Arts, no Letters; no Society; and which is worst of all, continued fear, and danger of violent death"; and, in contrast, "the Rousseauian picture of an African golden age of perfect liberty, equality and fraternity" (Hodgkin 1957, 174–175; quoted in Mudimbe 1988, 1). These two myths or white fantasies, are, of course, related.

In thinking about the first encounters—in the fifteenth century—of Europeans and Africans, Thomas DiPiero invokes the notion of trauma, but quickly backs off. He does not think that Europeans who traveled to Africa were somehow traumatized by the people they saw, except as their vision created what they "saw" there. ("It is said," Richard Dyer [1997, 209] reports, "that when sub-Saharan Africans first saw Europeans, they took them for dead people, for living cadavers.") "One creates the differences one sees," DiPiero (2002, 55) allows, "but the creation of these differences—that is, their *cause*—is lost in a signifying network that may, in fact, condition its own existence." We are caught, he suggests, in an "epistemological trap" that renders the origin invisible. The primal scene becomes intelligible retroactively, as we "regress," back to the tent.

In her discussion of Lacan's notion of *Das Ding*, Kaja Silverman provides a clue to Europeans' psycho–cultural predisposition to enslave Africans. "We confer . . . beauty," she writes, "when we allow other people and things to incarnate the impossible nonobject of desire [*Das Ding*]—when we permit them to embody what is itself without body, to make visible what is itself invisible" (2000, 17). Is that what Africans were to Europeans, the nonobject of desire, so beautiful as to be threatening: they had to be thrown out of the tent? Europeans did not "hesitate to lay violent hands upon other beings in this way; [Africans were] mere substitutes and surrogates, pale copies of eternal beauty" (Silverman 2000, 17). "Pale copies of eternal beauty" were precisely what Europeans, as Winckelmann's influence suggests, could tolerate; it was black embodiments of beauty that triggered the trauma of the Noah complex.

Such trauma operated in the life and work of Olaudah Equiano, the author of the first internationally best-selling African American autobiography and the prototype, Henry Louis Gates, Jr. (1988) suggests, of the nineteenth-century slave narrative. First published in London in 1789, the narrative of the *Life of Olaudah Equiano, or Gustavus Vassa, the African. Written by Himself*, makes it quite clear, Hortense Spillers (1987, 69) writes, "that the first Europeans Equiano observed on what is now Nigerian soil were as unreal for him as he and others must have been for the European captors." The cruelty of "these white men with horrible looks, red faces, and long hair," of these "spirits," as the narrator describes it, takes up several pages of Equiano's narrative, alongside a first-hand account of Nigerian interior life (quoted passages in Spillers 1987, 69). It is an interior life Europeans, in their sexualization of the black body, will not "see."

Converted to Christianity and to Calvinism specifically, Equiano confronted the binaries characteristic of Christianity and its Calvinist configuration: between good and evil, the elect and the reprobate. Transcending these was, presumably, the first step toward the accomplishment of Christian manliness. While white evangelists could "transcend" these oppositions (John Dewey [1960, 10] would term them an "inward laceration") by incorporating them into their savage, that is, enlarged, exalted (white male) selves, Equiano had far more difficulty overcoming these binaries, precisely because he was "black" and not fully, in white eyes, a "man." Equiano's struggle exposes, Carolyn Haynes (1998, 27) suggests, "some of the more insidious elitism embedded in Calvinist-based evangelicalism: its mission to annihilate and then incorporate the individual self, its cultural superiority, its propagation of a singular religious truth, and its support of a proslavery view of providence." Equiano's struggle with the Christianity he encountered underscores, Haynes (1998, 27) continues, "the limitations of a religious system based on hierarchical, binary relational structures where the goal is to augment one's own force, authority, or influence as well as to exercise dominion over others." Such religion rationalized the rape and enslavement of Africa.

Rather than authorizing a cultural order characterized by reciprocity and respect for the equality of differing subjects, Calvinism—and Christianity more broadly—constructed a subject-object relation wherein a "passive object serves as a vessel for God's and other authorities' projections and transferences" (Haynes 1998, 27). Conversion from the passive object role—vulnerable to sin and coded female and black—was through "the impersonation of the dominating subject" (Haynes 1998, 27), coded masculine and white. To be "saved," then, meant disavowing the female, embracing—through impersonation—God-the-Father, colored white, gendered male, and "victorious" not only over "sin" but the "uncivilized." But, as we have seen, this gendered and racialized religious system unravels, and the "feminine race" would become, by the late nineteenth century in America, the masculine race.

The racialized and gendered slippage between longing for/impersonating the God-the-Father and desire for/to be his bride/Virgin Mary has already occurred, in the garden, where God-the-Father has split his creation—his son

Adam—in "two." Rather than the Father, son, and the Holy Ghost, here we have the Father, his son Adam, and the son's rib which, split off from the male body, becomes "woman." DiPiero (2002, 216) thinks that unless we "exclude" the father from those social structures engendering men as "culturally male," we "face" the "taxonomically impossible task" of employing masculinity to "define masculinity." But that is exactly what happens in Genesis. God-the-Father differentiates among men by seeing who is tempted by the phallic serpent; he becomes s/he. Men's self-sexual differentiation is denied, then projected as lack onto "women," who are then commodified as sexualized property.

Drawing on the work of Cheryl I. Harris, DiPiero (2002, 195) considers whiteness as a kind of "property," that is, as a right, and not a thing. In this sense, being white "equates" to owning "property" (DiPiero 2002, 196), not only a reference to centuries of legal practice but to the ancient Israelite concern for genealogy, for succession, family as property, that is, patriarchy and power. We might think of white men's exteriorizations and racial-sexual projections as "lost property" that required, they felt sure, reclaiming. If the Western sense of self hinges on property ownership, and the exhibition of property, it is, Rey Chow (2002, 111) points out, "objecthood, rather than subjecthood, that defines the self," an elaboration of what she terms (see 2002, 112) the problematic of the "protestant ethnic."

DiPiero (2002, 197) focuses on men's experience of the disparity between the ideals of masculinity and their performance of it, engendering an "anxiety of insufficiency." He terms this anxiety "white masculine hysteria" (2002, 197). In the late nineteenth-century United States, "neurasthenia" was employed to distinguish men's hysteria from women's. As DiPiero understands, men's "crisis of masculinity" (a phrase not employed by DiPiero) follows from men's "fear of not coinciding with cultural ideals, and on the projection of and consequently identification with a lack in the other" (DiPiero 2002, 196). Since the fantasy of white masculinity is unattainable, DiPiero (2002, 196) argues, it remains an "alienated identity," conflated with other concerns, among them "economic" and "political property."[3]

Perhaps DiPiero's focus upon economic and political, rather than sexual, property, disables him from appreciating the landmark contributions of Trudier Harris and Robyn Wiegman, suggesting that while their work "accounts" for the white men's efforts to "feminize" black men, it does not explain "specifically" why they would "attempt to do so" (DiPiero 2002, 199). Certainly the path-breaking work of Harris and, especially, Wiegman shows us where to look: gender and sexuality. Inspired by their work, I argued that Southern white men's attempts to castrate black men derived from the gendered character of their defeat in the Civil War. The "anxiety of insufficiency" such men experienced was not only economic and political, it was specifically gendered, sexualized and racialized. "Insufficiency"—the disparity between reality and the ideal—was experienced as "lack," sexual lack, and more specifically still, as feminization. The limitation in Harris' groundbreaking work is its heterosexism. Wiegman (1993, n. 458) understood exactly that castration is an "inverted sexual encounter" between black and white men.

Late nineteenth-century Southern white men "regressed" developmentally in the face of the dissolution of that investiture (see Santner 1996, 124) with which they had earlier identified, not the least of which was the enslaved black body as white property. DiPiero (2002, 200) is disinclined to take his own clue when he notes that members of the Ku Klux Klan were "extremely fond" of "juvenile behavior." Psychoanalytically, adolescence is the developmental phase when the negative oedipal complex is typically (but not inevitably) resolved (see, for instance, Young-Bruehl 1996). Southern white men would seem to have regressed from whatever adult status they attained (on the backs of black slaves) in the antebellum period, to juvenile behavior, in which potency of adult black men (hallucinated as black male rape of white women) triggered that collective psychotic conduct known as lynching. DiPiero (2002, 205) cannot go there: "I am not suggesting that members of the KKK set out in high drag on their missions of terror, or that they had anything more in mind when they devised their disguises than making sure that they would not readily identifiable." One can devise numerous disguises to preserve anonymity; the question is precisely why *those* disguises?

DiPiero (2002, 205) relies on the notion of hysteria, noting that where self-division is denied, a "hysterical discourse" is given "voice." While the master—the hysterical postbellum Southern white man—believed himself, indeed, required himself to be "univocal," a certain kernel of repressed truth was expressed. DiPiero (2002, 205) writes that Ku Klux Klan members who imagined themselves as "100 per cent" men and as "uncontaminated whites" derived that fantasy from white women, by "project[ing] onto her body anxieties of gendered insufficiency or the possibility for racial contamination." I put the matter more bluntly: shattered by the Civil War, Southern white men regressed to earlier moments of convoluted identification with white women, wherein their negative complex was reactivated. Their fear of racial "contamination" through penetration by the black phallus was the disavowed desire for that very penetration. The master wanted to become the sexual slave he had once owned.

If racism is structured around gendered insufficiency in the Western white psyche, contradicted by fantasies of potency and property, we must recognize racism as a pathetic, if nightmarish, conversion of lack into fullness, debt into capital, the son's servitude into patriarchal possession and property. Racism is, then, a denial of castration. And castration, as Lacan (quoted by DiPiero: see 2002, 176) "can only be classified in the category of symbolic debt." In the late nineteenth century in the United States, that debt is not only symbolic. Reparations are owed to the descendents of former slaves on whose backs Southern whites built their economic—symbolic—prosperity. Interest has accrued.

By relying on the bodies of Africans in America to build his convoluted version of his repudiated paternal Europe, the Southern white man became bankrupt, propped up by false possession, everything but illusion lost in the Civil War. The repudiated father the son struggled to replace haunts the devastated postbellum landscape. Defeated, the clan of white sons dressed in women's clothes, in the color they fantasized white women represented: they sought those symbols of potency they no longer possessed. Now imagined as threatening their vulnerable white ladies—not real women, but imaginary elements of their own convoluted and devastated psychic terrain—black men became, in the white male mind, the stud son who must be banished from the tent.

"Man, disobedient to God, feels his disobedience in his very members," wrote the German Renaissance jurist-theologian Konrad Braun (see Steinberg 1996 [1983], 196). Not only Braun, but Jerome and Augustine before him, were speaking specifically, Steinberg tells us, of sexual excitation. If insurrection is sexualized, the covenant between an imaginary white father and the obsequious servile sons must be reasserted. The covenant is the lost cause that reverberates through the centuries.

Like those mythic sublimated sons—Japheth and Sham—the "heterosexual male position entails accepting castration," DiPiero (2002, 176), observes, the mark of which is circumcision. It entails as well, he continues, "acceptance" of a "particular order," understood both as an "arrangement" and an "injunction" (DiPiero 2002, 176). That "injunction" is nothing less than Western patriarchy, a racialized and gendered "arrangement" in which "others" become cursed as genealogical possession and "property." It derives from a "particular source," DiPiero (2002, 176) continues, soundly appropriately biblical. Once again he quotes Lacan:

> The assumption of the very sign of the virile position, of masculine heterosexuality, implies castration from the very beginning. That is what the Freudian notion of the Oedipal teaches us. Precisely because the male, exactly opposite to the feminine position, perfectly possesses a natural appendage, because he holds the penis as belonging, it's necessary that he have it from someone else, in this relation to what is the real in the symbolic—he who is really the father. (Quoted in DiPiero 2002, 176)

As I have suggested, the very notion of "opposite"—from the Genesis of the Western white imagination—represents a disavowal, a self-splitting of self-same desire, converting desire for the Father into desire for one's "opposite," fashioned from one's very inner structure, one's rib. Without inner structure, Western white men inflate themselves with fantasies of being their Father's only begotten son, impaled on a cross, the savior of the world.

In an otherwise brilliant book, DiPiero (2002, 178) founders on the phallus. DiPiero (2002, 175) complains that the isolation of sex as "fundamental" or in some way "originary" not only isolates sex from other structures of "difference," but simultaneously constructs a "unified field" of "meaning" against which all other structures of meaning are judged. This particular fantasy is one white men created. It is mythically installed in the creation of the species, and it appears, in Genesis 9, to have structured the founding moment of "race" in the white male imaginary. Sexual difference between men and women is not originary; it is itself derived from sexual difference within the male. Racial difference is derived from sexual difference in the male and, specifically, between father and son, castration restructured as servitude, self-mutilation transposed to the abjection of the other. Racism is the curse of the covenant.

That self-same male sexual difference gets coded "feminine" and has everything to do with the phallus, as men have desired and disavowed it. Surely DiPiero (2002, 181) is right to complain that we "get the phallus wrong" when we ignore "other ideologically charged bodily functions that participate along with sexual difference to graft signification onto the body," but the point is that these functions—orality and anality prominent among them—get signified after the originary moment of sexualized phallicized difference, when "man" is split into two in Genesis 9:21–22 by God-the-Father. The sons are split into two, the phallic Ham and his circumcised castrated brothers Shem and Japheth.

Does DiPiero (2002, 160) have in mind the rib when he asserts that the father "abstracts" the "mother's desire"? Can father even "know" the mother's desire in order to "abstract" it? Does not the concept of "mother's desire" risk being a patriarchal abstraction from the outset, even if "she" has told him what it is? Is not the abstraction DiPiero knows the reappropriation of men's desire displaced (in their own imaginary) as "mother's desire," making reproduction a matter between men in which mothers are consigned to "carry" the fetus, idea, culture that men create?

The "principal function" of the "phallus," DiPiero (2002, 159–160) writes, is the representation of desire in a "culturally contingent" and "coherent system of signification" structured through a "fundamental founding gesture." That founding gesture is the creation story, the fantasy of a self-made God, who requires of men that they should worship him absolutely. Tom F. Driver (1996, 56) suggests there is a necessary connection between narcissism and the idea of an absolute God. The reproduction of God-the-Father in Adam, the son's subsequent self-splitting, and the disavowal of desire—in the name of God-the-Father—signifies the gesture in which Adam's desire for the Father is severed and redirected, reified into a rib and made into "Eve," more accurately named, given the narcissism of the event, (St)Eve. If Noah is the second Adam, is not Ham in the same structural position as (St)Eve, now under the spell of the serpent?

As a separate ("opposite") creature conceived as his property, the concept "woman" was construed to be a "second sex," men's own "second sex." Women's historical emergence from positions men have required of them have threatened men's very self-structure, and for good reason: in men's minds "she" has always been an extension of that structure. Like Cleaver's Supermasculine Menial, however, "she" signifies the embodiment he cannot bear to experience in himself, unless, of course, he shatters, regressing to his preoedipal, "feminized" position. (For Cleaver, white men—the Omnipotent Administrators—were, in effect, so regressed.) DiPiero points to the abstract and inferential character of paternity, that, until very recently, rendered proof of paternity—that is, a direct empirical link uniting father and child—impossible. As Madeleine Grumet (see 1988, 10), has pointed out, the father could only claim his offspring "symbolically" and "inferentially": his link to the family is thereby "ideological, not biological" (see DiPiero 2002, 160). Paternity is, then, derivative, not direct: derivative from what, from whom? What is missing from view? (See Grumet 1988, 10).

DiPiero (2002, 160) recalls that Lacan's claim that the figure of the father was the "central obsession" governing Freud's study of the unconscious. Like Ham, Freud saw his father "naked" when his father declined to defend himself during that infamous anti-Semitic episode during Freud's youth. Lacan himself seemed to share this view of the unseen, arguing that "the most hidden relationship, and as Freud says the least natural, the most purely symbolic, is the relationship between father and son" (quoted in DiPiero 2002, 160). Precisely because the "negative" oedipal complex is repressed, the father–son bond is "hidden," including "behind" race. It "returns," however, in "symbolic recompense," social networks of sexual and racial "privilege" (2002, 161). Repressed, desire for the father becomes sublimated as hierarchical homosociality, those patterns of patriarchy in which the "traffic in women" is conducted, women as units of currency in a substitutional male sex economy. Just as white men obsessed over kinship and the possession of women, they came to desire and possess the bodies of black men.

THE MAN WHO CUTS

Since we never confront an object of observation directly, but must always pass through the lens of representation, any observation is necessarily culturally skewed. (Thomas DiPiero 2002, 48)

The lack of any religious affiliation of secular perpetrators of violence, however, should not make us ignore the religious character of collective violence. (Margarita Palacios 2004, 286)

A cut is capable of changing the very fabric of the signifying structure. (Levi R. Bryant 2004, 335)

During the late fifteenth century, Europeans discovered that tribes in what for them were remote and exotic parts of the world—Africa, the Americas, Australia, and Indonesia—performed a wide variety of circumcision-like surgeries on both males and females. In men, these procedures ranged from nicking or trimming off just the tip of the foreskin to a disfiguring mutilation that involved cutting the underside of the penis through the urethra all the way from the meatus to the scrotum. European invaders first imagined that these "primitive" peoples must share a common ancestry with the Jews. If, as the Bible taught, the origins of humankind could be traced from Adam through Noah and his sons, then even "remote" tribes had inherited circumcision from some ancient biblical patriarch (Gollaher 2000). Some believed there were "lost tribes" of Israel; in a book called *Jews in America* (1660),

Thomas Thorowgood felt certain that "many Indian Nations are of Judaicall race, seeing this frequent and constant Character of Circumcision, so singularlie fixed to the Jews, is to be found among them" (quoted in Gollaher 2000, 54).

By the end of the eighteenth century, however, as anthropology began to systematize European observations of "primitive" peoples, the tendency to link every circumcising tribe back to ancient Israel disappeared. That left unexplained the apparently universal practice. If primitive peoples had not inherited circumcision from Israel, from where did they inherit it? What was its meaning? If the mutilation of the penis was not a corrupt version of God's covenant with Abraham, what was it?

By mid-nineteenth century, European scholars began to formalize their investigations of "exotic" cultures. Sir Richard Burton was one of the most prominent of these; he was especially interested in sexual practices. Gollaher reports that in 1853 he had himself circumcised in order to pass for a Muslim when he traveled to Mecca. Burton wrote forty-three volumes describing his explorations among tribal peoples in India, Africa, and the Americas. He was hardly alone. In the study of circumcision, the pioneer was Arnold van Gennep, a young Belgian scholar who wrote a behavioral analysis of initiation rituals, *Les Rites de Passage* (1909).

Van Gennep suggested that male and female circumcision each sought to sharpen the distinctiveness of the sex organs by excising those parts—prepuce, clitoris, labia—that bore some resemblance to the opposite sex. With the glans permanently exposed, the penis appears to be in a state of permanent erection, a point not lost on the ancient Greeks, who considered circumcision indecent. Gollaher (2000, 57) suggests that the circumcised penis "may look more masculine." Those who practice female genital cutting often say that an uncircumcised woman, because her genitalia protrude (and thus resemble a miniature penis) appears not entirely feminine. Culture and biology evidently mix in the minds of some East African tribes who consider an uncut woman incapable of conception, or, should she conceive, of bearing a healthy baby (see Gollaher 2000, 58).

No tribal people have attracted more anthropological attention for their various genital mutilations than the aborigines of Australia. When the Dutch arrived in Australia in the early seventeenth century, as many as 700 tribes lived on the continent. After Europeans arrived in number (after 1788), aborigines died due to disease and economic exploitation (Gollaher 2000). With the emergence of anthropology in the late nineteenth century, the Aborigines came to be commodified as objects of study. Due to its isolation from modernizing and cross-cultural influences, Australia offered scholars an opportunity to study, as the pioneering anthropologist Baldwin Spencer phrased it, "human beings that still remain on the culture level of men of the Stone Age" (quoted in Gollaher 2000, 59).

Along with F.J. Gillen, Spencer traveled throughout the Outback in the 1890s. For the next thirty years the two men studied dozens of aboriginal tribes, particularly the Arunta. Their accounts, and those of Herbert Basedow, a German-trained physician and anthropologist who also served as

Australia's Chief Medical Inspector and Protector of Aborigines in the Northwest Territory during the 1920s, documented rites of passage and genital cutting as they existed in these "primitive" cultures. They found that rites varied greatly within regions and among tribes. Communities that did circumcise (or practice sub-incision) had created elaborate ceremonies. Although none could be said to be "typical," there were common elements, making possible the creation of a composite description (Gollaher 2000).

Not every tribe engages in circumcision, but among those who do, Gollaher (2000, 60) tells us, the practice is "public" and "charged" with "great significance." Deciding when a boy has reached the age for circumcision (usually around twelve years old) is left to the male members of his family. Their decision is withheld from him, and so, suddenly, without foreknowledge, the boy's brothers, or other older males who act as "designated brothers," abduct him. He is taken to a small outpost they have prepared some distance from the tribe's main encampment where he remains, confined and closely guarded. He is to have no contact with others, especially girls and women (Gollaher 2000).

Not only is the outpost isolated from his mother, any sisters or aunts, and from all other women, but the place where the ritual will take place is also off-limits to women. The men who will participate in the circumcision proceed to decorate their bodies with red and white down; others clear a space in the underbrush, loosen the dirt with sharp sticks, then cover the soil with leaves from a red gum tree. A huge bonfire is ignited. Dancers gather around the boy, circling the fire, singing, chanting, terrifying—Gollaher (2000, 60) characterizes him as the "victim"—by making ferocious faces. At an appropriate moment of intensity, the boy is taken to a second, smaller fire, some distance away. There men smear his naked body from head to foot with red ochre, truss his hair, then return him to the main site (Gollaher 2000).

At this point, when the boy is "disoriented" and "mortified" (Gollaher 2000, 60), the elders tell him tribal secrets, threatening to kill the terrified boy and his family should he ever reveal this information. In some tribes, this phase of the ritual takes several days. Possibly to add to the drama (and to his terror), the boy is sometimes blindfolded and unblindfolded as the elders speak to him. The "instruction" stops as the moment of cutting approaches, and dancing men crowd around the boy, seize him, then lift him up and carry him forward. Their shrieking and chanting become intense, and more and more hands grab at the terrified boy until, without warning, he is placed onto the prone bodies of men who have arranged their bodies to form a "human operating table." Lying on their backs, these men position the victim, holding his arms and legs, while another man sits on his chest to prevent his movement. Men who have undergone circumcision themselves dash about, brandishing burning sticks and screaming into the night (Gollaher 2000). A boy is about to become a "man."

To prevent him from screaming, one man stuffs hair-string material into the mouth. Those men holding his legs spread them and pull them downward, exposing his genitals. At this instant, the crowd parts and the man designated to cut the boy moves toward him. He has the appearance of a man possessed, "his beard between his lips and his eyes rolling in their sockets" (quoted in Gollaher 2000, 61). Using a knife chipped from flint or quartz, he begins the cutting. After several strikes, the prepuce is severed. The cutter holds the foreskin aloft to raucous approval and men lift the boy above their heads, allowing his blood to spill onto a piece of bark (Gollaher 2000).

The wound seems key to the ritual, although tribes differ in their treatment of it. Some northern tribes, Gollaher reports, cover it with thin sheets of bark, soil, emu fat, and hot ashes to stop the bleeding. Once the boy recovers, he is sometimes presented with a spear or a shield. These are, for the tribesmen, "badges" of "manhood" (Gollaher 2000, 62). During his recovery, the boy is kept in the bush apart from the tribe, fed a special diet, and closely watched for signs of sickness. When, at last, he returns to the tribe, he wears a fur tassel over his penis; his mother, sisters, and aunts wail, tearing their hair and pricking their bodies in sympathy for the suffering they sense he has endured (Gollaher 2000).

Many pre-initiates, fearing the psychological and physical pain of aboriginal circumcision, run away when they sense their time approaching. Some become desperate, begging anyone to hide them. Sometimes there is a sympathetic European who does hide the boy, but eventually most runaways are found and dragged back. As punishment and to set an example, tribal leaders makes these initiations "especially excruciating" (Gollaher 2000, 62). Climb onto the cross, my son, in the name of the Father.

The origins of aboriginal circumcision remain unknown. Replying on the research of anthropologist M.F. Ashley-Montagu, Gollaher reports that one South Australian tribe attributes the practice to a mythological creature called *Jurijurilja*. Legend holds that the primordial beast hurled a boomerang that flew back, flaying the foreskin of the man's penis and, in the same motion, passing through the genitals of his wives, thereby linking circumcision and menstrual bleeding (Gollaher 2000). We are back with Bettelheim.

"Years ago, when I learned of sub-incision—the remarkable practice of cutting the ventral portion of the penile urethra, sometimes from the glans to the scrotum—I puzzled over its meaning," Ashley-Montagu admitted, "until I found that, among the Errand of Central Australia, the sub-incised penis was called by the same name as the female vulva." Sub-incision was, evidently, designed to reshape the penis in the image of the vulva. The subsequent hemorrhage was likened to menstruation by means of which women were able to dispose of the evil accumulating in their bodies. "To continue the same effect, males periodically engaged in incision of the penis and called it menstruation" (quoted passages in Gollaher 2000, 63). Do boys just want to be girls? Are they already?

Perhaps the most "meticulous" examination and theorization of a circumcision ritual in a historical, social, and cultural context, Gollaher (2000, 63) judges, is Maurice Bloch's fifteen-year study (during the 1960s and 1970s) of the Merina of Madagascar. The Merina circumcise between the ages of one and two years. The idea that a boy would not be circumcised is "inconceivable," although villagers were unable to explain why they felt so strongly. Some referred to tradition while others believed that the ritual made boys "sweet" or "beautiful" and "clean" (Gollaher 2000, 63). Others felt certain that, without circumcision, boys could not become men, would never achieve sexual potency. The most deep-seated belief was that circumcision constituted a blessing. In Merina culture, blessing would to be, as in North American terms, a "spiritual" concept involving some sense of the boy's destiny and his significance to kin and the larger community (Gollaher 2000; Bloch 1986). As we have seen, the sexual and the spiritual are not separable; neither, it seems, can we distinguish between "blessing" and "curse."

There are many involved in the dawn ceremony. Primary, perhaps, is the man who cuts, a man who, despite being unrelated, is called "father of the child" (quoted in Gollaher 2000, 64). This conflation of "father" with the one who wounds is reminiscent of the monotheistic God-the-Father. Is incorporation—the Eucharist in Christian traditions—also a sexualized act? Freud's theorization of the oral as a primary and formative stage in psychosexual development suggests so, that is, that incorporation has sexual connotations, even when not involving genitalia. Among the Merina circumcision does involve genitalia. As soon as the young boy's penis is cut, the circumciser hands the child's prepuce to an older male relative, who sandwiches it in a small piece of banana and eats it (Gollaher 2000, Bloch 1968). Eat this in remembrance of me.

Relying on Bloch, Gollaher locates this act of incorporation culturally rather than sexually, suggesting—no doubt accurately, as far as this explanation goes—that in Madagascar, as elsewhere, the foreskin has talismanic significance. (As we saw earlier, the talismanic significance of Christ's foreskin condemned it to circulate vampirically, unable to rest.) Its disposal is suggestive of its meaning, Gollaher tells us, but he does not pursue this point. Instead, he notes that many tribes bury foreskin, sometimes in dry soil, sometimes covered by soil bloodied during the operation, and sometimes in an anthill where it may be incorporated by the earth and thereby protected from evil. There are, Gollaher continues, aboriginal tribes in Australia who hide dried foreskins in secret spots invested with sacred significance: rocks, hollow trees, caves, and other totemic places. In other tribes, the prepuce is presented to a sister of the initiate; she dries it, colors it with red ochre and wears it on a string necklace. During the 1920s, among the Ait Yusi in Morocco, the foreskin was presented to the boy's mother, who attached it to a little stick taken from her spindle. Afterward, she hung the prepuce over her family's tent for a week; then she discarded it (Gollaher 2000).

Eighteenth-century French naturalist and historian Georges de Buffon's report that Persian women swallowed their son's foreskins to ensure fertility seems even more explicitly sexual. Some mothers preferred their foreskin cooked, however, and several Australian tribes roast the severed prepuce over a fire before presenting it to her. The Hova prefer to give the severed skin to the circumcised boy's father, but if he declines, it is wrapped inside a banana leaf and fed to a calf. In Mali, there are Dogon tribes who grind severed fore-skins with millet, making small cakes that are consumed by the circumcised boys themselves on the third day after the ritual (Gollaher 2000). Does this enable the boy's penis to rise from the dead? Early Renaissance Christian painting, as Leo Steinberg (1996 [1983], 81) reports, featured an "erection" on the "figure of the dead Christ." He is risen.

DON'T LOOK, JUST TELL

The aversion of the gaze, I've suggested, also reflects an ambivalence about God's sex. (Howard Eilberg-Schwartz 1994, 77)

[W]hite women become silent markers in the systems of exchange that make both whiteness and heterosexuality cultural givens. (Mason Stokes 2001, 17)

[T]he relationship between castration and racism is not accidental or anecdotal or limited to the American South, but fundamental and structural. (Gary Taylor 2002, 16)

The gaze averted—to which Howard Eilberg-Schwartz refers in the sentence quoted above—is that upon God-the-Father, recalling another gaze upon another father, Noah. Inside Noah's tent, according to one stream of biblical exegesis, the son's visual attention to his naked father provoked his father's curse. It was, in this tradition of commentary, the "sin" of Ham. It punishment was servitude, which Christians later understood in racialized terms.

For some, Eilberg-Schwartz acknowledges, the aversion of the gaze from the deity's face (Exodus 33)—the eyes are deflected to the feet (Exodus 24)—is self-evident; it is, apparently, an act of deference. But other cultural currents may well be embedded in this simple, seemingly self-evident, act. As he argued earlier (Eilberg-Schwartz 1990), culture is a palimpsest of meanings: layers of meanings are superimposed one upon the other. In fact, myths and rituals become powerful "precisely" when they convey multiple messages "simultaneously" (Eilberg-Schwartz 1990, 119–140; 1994, 77). The deflection of the gaze to the feet, he suggests, conveys more than one meaning.

First, Eilberg-Schwartz points out that by looking to the feet one is able to avoid looking at God's midsection. But to the faint of heart there is no safety in feet, as Eilberg-Schwartz observes that the term "feet" is an occasional euphemism for penis (Isa. 7:20; Ruth 3:7; possibly Exod. 4:25). Moreover, covering the feet is apparently a euphemism for urination (Judg. 3:24; 1 Sam. 24:4). Clearly, in ancient Israel an aversion to referring to the penis is performed by deflecting attention from it onto the body's extremities. It is in this context that Eilberg-Schwartz discusses Ezekiel's encounter with God.

Ezekiel has, Eilberg-Schwartz points out, a direct frontal view of the deity, but focuses his gaze on God's loins. His description proceeds from the loins down, rather than from one end to another, as if his eyes were drawn to the midsection of the deity's body. If Ezekiel's description is in any way suggestive, Eilberg-Schwartz notes, it is the deity's midsection that stood out in a frontal view. The erotic overtones of this vision played an important role in esoteric doctrine in early rabbinic mysticism, as Eilberg-Schwartz discusses later. The effort to de-eroticize looking would seem to typify certain currents in the ontology and epistemology of vision in the West, as we have seen.

Eilberg-Schwartz notes that Genesis 9:20–25 has several striking similarities with the story in which God turns his back to Moses (Exod. 33:12 ff.). God turns away so Moses cannot see the body of the Father, just as Shem and Japheth walk backwards to avoid seeing their father's genitals. Moreover, the father's nakedness represents other acts that dishonor the father. In the Holiness Code (Lev. 17:1–26:46), Eilberg-Schwartz explains, incest with one's mother is on several occasions described as uncovering the father's nakedness (Lev. 18:7–8, 20:11) and, in language reminiscent of the Noah passage: "Your father's nakedness, [that is] the nakedness of your mother, you shall not uncover; she is your mother—you shall not uncover his nakedness." The prophet Ezekiel (22:10) also speaks of the sin of "uncovering one's father's nakedness," when referring to incest between a son and his mother (Eilberg-Schwartz 1994).

In this homosocial economy, sex with one's father's wife constitutes appropriating his "property," a sentient extension of his own self, as would be the slave an economic, psychological, and sexual extension of the slavemaster (Hartman 1997). In such a gendered economy, then, sex with the mother is, in effect, sex with the father. After all, as the Genesis creation myth asserts, the "woman" is his "rib," a part of "him." Who she is—apart from his fantasy of her—he has no clue. Entering his tent—a metaphor for the feminine in ancient Israelite culture (see Boyarin 1997)—suggests entering his inner nature, invoking his vulnerability, stimulating his "unmanning," his "becoming-woman."

Eilberg-Schwartz (1994, 82) points out that in the Holiness Code "uncovering the father's nakedness" is no euphemism. The phrase does not replace but supplements the act of "uncovering the mother's nakedness." Why would incest between a son and his mother, Eilberg-Schwartz (see 1994, 82) asks, be recoded as "uncovering his father's nakedness"? He argues that interpreters who take the "father's nakedness" as being (only) an indirect reference to the woman's nakedness, fail to acknowledge that the father's nakedness is in itself dishonorable. Eilberg-Schwartz cites the Noah incident as precedent: clearly, he notes, it is disgraceful for a father's nakedness to be exposed, an assumption not limited to one author or one period of time but commonplace in ancient Israel. The Noah incident suggests to Eilberg-Schwartz that the taboo against looking at the father's body—an act which compromises him by "unmanning" or "castrating" him—lies behind the tale of God's turning his back to Moses, even if that myth does not, evidently, derive from the same tradition as the myth of Noah (see Eilberg-Schwartz 1994).

The invisibility of the divine father's body, Eilberg-Schwartz (1994, 85) argues, is "analogous" to the clothing of the human father's nakedness in the story of Noah. What neither Israel nor Noah's virtuous sons (Shem and Japheth) dared see was the penis of their father. This cultural prohibition against what Freud would much later term the "negative" oedipal complex might well be unique to ancient Israelite culture. As I have as extract, Eilberg-Schwartz cites the work of Leo Steinberg (1983), scholarship that reveals how the penis of Jesus was depicted as erect in medieval and early-Renaissance art, presumably an expression of Jesus' humanness: "Nakedness becomes the badge of the human condition which the Incarnation espoused" (Steinberg 1996 [1983], 33). It would also seem to expresses Jesus' maleness and, perhaps, a dissociated homosexual desire embedded in the worship of a nearly naked man hung on a cross. "Following Steinberg," Richard Dyer (1997, 68) points out, "if Christ's humanity was to be fully depicted, then not only must the difference of his sex be represented but so also, in an age of increasing 'racial' awareness, must the difference of skin color."[4]

As does David Gollaher, Eilberg-Schwartz notes that the foreskin of Jesus' penis became a sacred relic, even a fetishized object of veneration in medieval Christianity (recalling that at least one medieval nun believed the foreskin of Jesus was used as a wedding ring in her marriage to Christ). But medieval Christians have hardly been the only devotees of the divine to fasten upon the male genitals; Eilberg-Schwartz tells us that early Buddhist sources describe Buddha's body as having seven marks, including a retractable penis (like that of a horse).

While only the Buddha may have been hung like a horse, other semi-human, semi-divine men—Eilberg-Schwartz is thinking of the Greek gods Poseidon, Apollo, and Zeus—are frequently portrayed with their penises fully protruding. There is the Greek myth of the God Ouranos, the God of the sky, who was castrated by his son Kronos. Was Kronos, like Ham's son and descendents, enslaved in return? No, Ouranos' manhood falls into the seas and from it Aphrodite is born. Zeus's sexual exploits are narrated in many Greek myths as well (Eilberg-Schwartz 1994).

The erect phallus of Siva is the subject of Hindu mythology, a symbol of the power to spill the seed as well as to retain it. Hindu mythology tells a tale of the wives of Pine Forest sages who touch Silva's erection. It tells also tells the tale in which a woman finds a penis and, thinking it to be Siva's *linga*, takes it home and worships it. At night she uses it to stimulate herself sexually. A Hindu textbook on aesthetics directs students to draw Siva riding on a bull with an erect penis, the tip of which must reach the limit of the navel. In still another myth, Siva castrates himself because there is no use for this *linga* except to father beings who have already been created (Eilberg-Schwartz 1994).

The divine phallus, Eilberg-Schwartz reports, shows up in Near Eastern mythology. There are Sumerian stories that describe how the God Enki masturbates, ejaculates, and fills up the Tigris with flowing water. Then he uses his penis to dig irrigation ditches, a key feature in the Sumerian agricultural system. The sexual exploits and erections of El are the subjects of religious poetry. As these myths make clear, Eilberg-Schwartz (1994, 856) points out, the "divine phallus" has not been suppressed in other traditions. Why, he asks, in ancient Israel?

To answer that question, Eilberg-Schwartz turns to the nakedness of Noah. He notes, as have Cohen and other biblical exegetes, that the Genesis passages do not reveal why it was such a sin for his son to gaze upon his nakedness. The writer takes the prohibition as something taken-for-granted. Eilberg-Schwartz reminds us that several ancient and modern commentators have concluded that Ham sexually assaulted his father, or, possibly, castrated his father. This argument tends to turn on the fact that when a son committed incest with his mother it was referred to as "uncovering the father's nakedness" (Lev. 18:22). Possessing the father's "property" dishonored him, we might say "castrated" him. "Uncovering the father's nakedness," these commentators conclude, must refer to homosexual incest in the Noah story (Eilberg-Schwartz 1994).

The laws of incest in the Holiness Code, Eilberg-Schwartz reports, underscore how son–mother incest, like the father's indecent exposure, dishonors the father and discredits the patriarchal regime. Eilberg-Schwartz is skeptical of the homosexual assault hypothesis, pointing out that had that occurred, why did the other brothers walk backwards and turn their heads? He points out that this behavior only makes sense if looking upon the father's nakedness was the sin. "The myth of Noah's nakedness," Eilberg-Schwartz (1994, 87) concludes, "therefore makes most sense as a condemnation of the father's exposure before his sons." Although he may well be right (and the rape theory wrong), I suggest that it does not, finally, matter, since the gaze in ancient Israelite culture amounts to a kind of "rape," castrating in its impact and gendered (and, later, racialized) significance.

Eilberg-Schwartz focuses on the impact of the gaze within the patriarchal—homosocial—culture of the ancient Jews. Within that context, the patriarch's "honor" was compromised. It was, Eilberg-Schwartz (1994, 87) suggests, "disrespectful" for Ham to see Noah naked, especially given his vulnerable drunken state: "To avert the gaze was to respect the father's honor." Noah's condition is important in this interpretation, Eilberg-Schwartz continues, because in Israelite imagination, the father's nakedness was connected with shame when the father was the passive object of someone's gaze. A father was not dishonored if he intentionally exposed his body: it was, Eilberg-Schwartz tells us, his prerogative to do so. I suggest that honor is the ritualized residue of homoeroticism in patriarchal culture; its brittle social structure bears witness to its explosive fragility, as its violation required, in recent centuries, a duel to the death. Among the ancient Israelites, evidently a curse would do.

Eilberg-Schwartz seems to accept "honor" at face value (rather than as a convoluted cover-up and ritualized regulation of men's desire for each other) and focuses instead upon Noah's drunken state, upon his incapacity to control the gaze of the son that is key to decoding the event. He points out that the narrator of the Noah story—the J source—is not timid in portraying a servant or son grasping a patriarch's penis while taking an oath, as evident in two other stories from the same source (Friedman 1987; Eilberg-Schwartz 1994). In the first (Gen. 24:1–4), Eilberg-Schwartz reminds, it is Abraham who orders his servant Eliezer:

> Put your hand under my thigh and I will make you swear by the Lord, the God of heaven and the God of the earth that you will not take a wife for my son from the daughters of the Canaanites among whom I dwell, but will go to the land of my birth and get a wife for my son Isaac . . . . So the servant put his hand under the thigh of his master Abraham and swore to him as bidden.

The servant's hand on his master's penis sealed the deal. When did the substitute ritual—the handshake—appear?

In the second story (Gen. 47:29–31), Jacob asks his son Joseph, "Do me this favor, place your hand under my thigh as a pledge of your steadfast loyalty: please do not bury me in Egypt." Joseph replies: "I will do as you have spoken." And Jacob says, "Swear to me." "And he swore to him." Eilberg-Schwartz points out that the word "thigh" is sometimes used in ancient Jewish sources as a euphemism for the penis. In both of these stories, he notes, the patriarch has exposed his nakedness intentionally in the assertion of his power and status and to confirm an obligation. "The penis is the symbol of the patrilineage itself," Eilberg-Schwartz (1994, 88) concludes, adding, "Noah, by contrast, has been viewed in a drunken stupor, a shame act." These sentences reiterate my point, that patriarchy—and its key elements, among them "honor" and allegiance and (genealogical) succession—represents, albeit in various forms according to culture and to historical moment, a sublimated regimen animated and socially sealed by what men feel for each other. These are boys' stories, testifying to their fantasies of the father, of what a "man" might be, God-the-Father phallus and the son punk bitch. The formation of sexual—and racial—identities is traumatic (see Cvetkovich 2003, 44–45)

Not only *la femme n'existe pas*, neither does "*l'homme*." After all, the penis is not the phallus. To experience the (male) self as negation is to go back inside Noah's tent, to experience the "negative" Oedipus complex in which sons desire their fathers and fathers their sons. Eilberg-Schwartz notes that there has been considerable attention paid to the taboo on heterosexual incest, but little attention to how the social prohibition on incest between sons and fathers developed. He is interested in how sons are taught to desire people like their mothers, but not like their fathers. He wants to know what role the repression of father–son incest plays in the accomplishment of compulsory heterosexuality. Does the latter depend upon the former?

Eilberg-Schwartz notes that Freud's acknowledgement that infant sexuality can take multiple forms undermines those who embrace a biologistic determinism, that sex is destiny. He turns to anthropological research as well, citing the work of Gilbert Herdt (1982) on the Sambia, work reviewed (through David Gilmore and Gerald Creed) earlier herein. The variable forms that sexuality assumes across cultures suggests, Eilberg-Schwartz notes, that culture plays a key role in the organization of desire. For him, anthropological research corroborates Freud arguments that desire is shaped by familial and cultural experience, themselves intersecting categories, to be sure.

Eilberg-Schwartz returns to the myth of Noah, emphasizing that the prohibition against the son seeing his father's nakedness symbolically expresses and, in fact, institutionalizes the prohibition against homosexual desire. "Heterosexuality" is made compulsory. The story of Noah is, Eilberg-Schwartz suggests, a second creation story, in psychoanalytic terms a "primal scene" in which the reproduction of the species—and the regime of hetero-sexual desire which makes it possible—is divinely ordained. Eilberg-Schwartz points out that in Genesis 2:23–24, a story from the same author as the myth of Noah (J), heterosexuality has already been ordained by God, who, we are told, has decided that "the earthling should not be alone" and takes a piece of Adam's side to create a second creature. This is also, in biblical terms, the founding of patriarchy, the idea that women exist to serve as man's complement. But this gendered arrangement—so integral to men's regulation and domination of reproduction—is, Eilberg-Schwartz observes, potentially disrupted by another erotic relationship: that between sons and fathers. For Eilberg-Schwartz (1994, 93), the "prohibition" against "seeing" Noah's nakedness follows from the danger of this disruption.

Eilberg-Schwartz's interpretation of the Noah myth is grounded in the prohibition against male–male sexual activity in Israelite culture, a prohibition evident in ancient texts. (Daniel Boyarin argues that the prohibition in ancient Israelite culture was against male–male anal intercourse only, a quali-fication that does not alter, I think, Eilberg-Schwartz's interpretation. It does alter our understanding of Leviticus.) The Holiness Code treats male–male sexual acts (again, Boyarin argues the abomination is anal intercourse only) as an abomination (Lev. 18:22), punished by death (Lev. 20:31). Eilberg-Schwartz finds the language of the code significant: "Do not lie with a man as one lies with a woman; it is an abhorrence." It is language, he suggests, that is consistent with Israelite efforts to preserve their perceptions of the "natural order." Perhaps the "natural order" is in the mind's eye and the ancient prohibition is only an expression of positional preference.

CHAPTER 8

The Sodomitical Subjectivity of Race

VISUALITY AND A MASTERING SUBJECTIVITY

The ego forms itself around the fantasy of a totalized and mastered body. (Elizabeth Grosz 1995, 86)

[I]t is human freedom that is undermined by the look of the Other. (Martin Jay 1993, 289)

Does changing the world require ending the hegemony of vision? (David Michael Levin 1993a, 23)

Both the ancient Israelite taboo against looking at the body of the father and God-the-Father's "pantoptical arrangement" (Sass 1994, 156 n. 45) with Schreber links visuality to an objectifying, mastering subjectivity, elucidated, Stephen Houlgate (1993) tells us, in the work of David Levin. For Levin, vision is the "most reifying" of our "perceptual modalities," that mode of perception which, more than any other, renders the world as objects, presumably present and ready-at-hand for our use (quoted in Houlgate 1993, 96). For Noah and Schreber, "reifying" is gendered: it implies "feminizing." Like Sass, however, Levin focuses on the epistemological point: vision creates the illusion that what is seen is there, as it is seen. Stripped of subjectivity, what is seen can, presumably, be surveyed and mastered: "For modernity, vision has become supervision," Thomas R. Flynn (1993, 281) asserts. For Levin, this is the "power drive inherent in vision," the tendency of total visibility toward total control over things (Houlgate 1993). As we have seen, this tendency is gendered and racialized.

In the West, the "hegemony of vision" in the modern world is associated, then, with racism and misogyny. Epistemologically, it rests upon the unquestioned cultural conviction that vision is the most compelling means of knowing. It is, for Levin, associated with those forms of power dominating contemporary social, economic, and political life, including the scientific and technological exploitation of the earth, as well as contemporary political surveillance (Houlgate 1993; Levin 1993). This analysis owes an obvious debt not only to Heidegger, but to Foucault's analysis of "biopower" as well, in which biological life in all its cultural forms is regulated by the state of surveillance, what Michael Hardt and Antonio Negri (2000, 208) term the "non-place" of "Empire."

After Heidegger, surveillance in the service of power is not the only function of visuality. For Levin, vision also enables us to focus upon what is there in order to bring it directly into view. Such visual acts can be "relaxed, playful, gentle, caring," more *gelassen*, as Heidegger would put it (quoted in Houlgate 1993, 96). Such looking is not necessarily "fixated" on the objects present before one's eyes, but, instead, is "diffused, spacious, open, alive with awareness and receptive to the presencing of the field as a whole," that is, a "gaze that does not seek to control things, but lets things show themselves as they are, and so lets them *be*" (quoted in Houlgate 1993, 97). These—either active control or passive acknowledgement of reality—would seem to be the only two options for the "father." In contrast, the "maternal" point of view can be engaged, relational, committed, occupying a space "in-between" control and acknowledgement, loving but giving the other, whom one "recognizes" as an independent subjectivity, freedom.

Unmindful, apparently, that these are mythically "feminine" adjectives, Houlgate (1993, 97) wonders if vision can be made more "subtle" and "supple." Is he not asking how a man's perception might be more like a woman's? Houlgate reports that Levin's answer is to attune vision to what is *not* directly before the eyes, to that which withdraws from the social surface and, thereby, is hidden from surveillance and control, a "secret place" perhaps (Langeveld 1983; Pinar et al. 1995, 440 ff.) Levin gestures toward "shadows" and "reflections" within fields of vision, not *objects*, Houlgate underscores, modalities of reality not directly seen (for example, a hidden source of light), and which "deepen, heighten, extend and enrich the field of visibility" (quoted in Houlgate 1993, 97).

A mode of vision attuned to shadows and horizons would not constitute a "stare, an act of direct, frontal looking fixated on its object," but, rather, "a playful gaze . . . which delights in ambiguities, uncertainties, shifting perspectives and shades of meaning." Does this not sound "gay"? Do *women* need to affirm ambiguity and uncertainty? If we were to base our knowledge on such envisioning, rather than on objectifying visuality, Levin claims, this grip of the will to power governing modern humanity—would he include patriarchy and racialized masculinity?—might well be loosened (see Houlgate 1993, 97). Would racism and misogyny then diminish or would they mutate into not immediately recognizable forms?

Vision is not intrinsically reductive and politically reactionary, Levin insists. Houlgate underscores this point, suggesting that a more open, responsive form of disclosure is not achieved by *seeing the objects before us more clearly or completely*, but, rather, by becoming attuned to what is not directly before our eyes, to "elusive horizons," "shadows," and "reflections" (Houlgate 1993, 97). I am reminded of gay men "cruising," watching for looks returned, a shadow of desire hidden underneath the social surface and, for some "straight" men, underneath their field of vision. Indeed, gay life has been described as "shadowy," as an "underworld." To be "opened up,"

Houlgate (1993, 97–98) suggests (as if taking his cue from queer theory), vision must be "taken out of itself" and become attuned to that which is not visible, to straight men's subjectivity folded into musculature. I am looking at your body but it is your subjectivity I also see. "Appearance," Silverman (2000, 3) asserts, "is the locus within which Being unfolds."

Houlgate questions Levin's analysis, asking if vision (and that mode of thinking which is modeled on vision) is inherently condemned to dominate that which it surveys. He wonders why the interest in clarity cannot also be a willingness to allow reality to disclose itself (see 1993, 98). If so, would that not mean that the interest in the clarity of vision (and that objectivity it presumably affords) is, rather than to blame for technological domination of the earth, that which holds out the prospect for loosening the grip of that drive? If it could be shown that objective vision and objective thinking are not necessarily reductive and politically reactionary but, instead, can support a genuine (including a gendered and racialized?) openness and attentiveness to world, Levin's critique—resting as it does on both the phenomenological and post-phenomenological project of deconstructing the "metaphysics of presence" and, thereby, undermining the "hegemony of vision" that is believed to have given rise to this metaphysics—would be called into question. My own view is that vision is not inherently anything. Ham's "looking" is cursed because he is the son who sees that Noah is Schreber in disguise.

Houlgate (see 1993, 111) argues that a particular kind of thinking, not vision itself, is responsible for reducing the world to a realm of objects. Yes, but it is a particular order of thinking that employs vision to manipulate and dominate. The "situatedness" of "consciousness" proves key (1993, 111). Reducing the woman to her embodied position in my desire and thereby effacing her subjectivity performs a particular politics of masculinity that devalues femininity *by* desiring it, by making it an object in specular male (it would operate differently if it were another woman's) desire, a desire that is propped up socially, politically, and economically, by religion and culture. For gay men, "cruising" functions differently politically, even when "straight" men are its "object." Because hegemonic masculinity cannot be reduced to a sexual object—a gay man's political, political, economic, gendered standing in society contradicts such a reduction—cruising cannot function politically as does the straight-male commodification of women's bodies.

What can be, in fact, politically progressive about cruising straight men is, in part, its potential to bring them down a notch. The firestorm set off by the 1993 presidential policy on gays in the military underscores how threatened some straight men are by the presence of gay men (see Bersani 1995 for a brilliant and amusing commentary on this event). But the politically progressive potential of the policy was not only unrealized, it stimulated political reaction, even violence, as subsequent murders of gay men in the military document. What is politically progressive about cruising straight men has to do with the look's potential to de-objectify and re-subjectify straight men. In many heteronormative men, subjectivity has been banished to body. Structuring their musculature is repressed emotion, including the residue of the masculinized repudiation of the preoedipal identification with the maternal body. That muscle also includes memory of the son's desire for the father. To bring homosexual attention to the surface of the (straight) male body threatens to de-crystallize that repression, threatens its resurfacing, as he, like Schreber, now capable of (be)hindsight, experiences himself as "unmanned," as "woman," now "voluptuous."[1]

Houlgate concludes his helpful essay by suggesting that the twentieth-century emphasis on the problematic nature of vision and the corresponding de-emphasis of the problematic nature of reflection leads us to overlook those reflexive features which lead to manipulation and domination. Both Dewey and Rorty succumbed to this problem, Houlgate (1993, 116) asserts; their replacement of the "spectator" theory of knowledge with a conception of knowledge based on the pragmatic idea of "coping" or "dealing" with the world perpetuates the "technological" assumption that our "primary" way of being-in-the-world is that of reconstructing the world "for" ourselves, rather than allowing the world to be disclosed as "whatever it is in itself." Houlgate may be exaggerating, although hardly fabricating, this difference between American pragmatism and European phenomenology.

If we gender and racialize the "spectator theory of knowledge," its resonance with "whiteness" and with hegemonic masculinity comes into focus. Invisible to itself, whiteness assumes that its observations of the "other" constitute "reality," when, in fact, it was seeing its dissociated "self" in its fantasies of black hypersexuality. Invisible to itself (or, in Lacanian terms, misrecognizing itself in its gendered and racialized mirror), hetero-normative masculinity assumes that its observations of "blacks," "women" and "gays" are self-evident. "To make something of the world for ourselves" can be, for whites, an autobiographical project, as it is precisely "something of the world" that has, historically, been split off and subjugated by white racism and the "separate-spheres" ideology. Aimed at dismantling masculine superstructure, Bersani's self-shattering is a more dramatic version of Houlgate's (and Heidegger's) "letting it be," meditatively allowing the substructure of heteronormative masculinity to surface. "Becoming-woman," Noah morphs into Schreber.

Rather than denying one's own self-same alterity through self-dissociated specularity, rather than denying one's own "woman-identified" interiority, let us rewrite Genesis 9:24. In this proclamation of emancipation, Noah and son embrace, each allowing himself to become the object and subject of desire, a son, simultaneously the father, not a panoptical patriarch or omnipresent holy spirit, but an enfleshed self-differentiating trinity inside the tent, that primal scene of the sodomitical subjectivity of men in the West. There, as Elizabeth Grosz (1995, 109) writes:

> The subject ceases to be a subject, giving way to pulsations, gyrations, flux, secretions, swellings, processes over which it can exert no control and to which it only wants to succumb. Its borders blur, seep, so that, for a while at least, it is no longer clear where one organ, body, or subject stops and another begins.

If no longer men, no longer white, can the curse then be dispelled? Or is this only another (gay) white man's fantasy?

FANTASY AND HISTORY

Homosociality, fueled by hatred but more often than not rendered as same-sex eroticism, is ultimately what makes white supremacy possible. (Mason Stokes 2001, 137)

Psychohistories of white racism have always called attention to the tension between the construction of the black male body as danger and the underlying eroticization of that threat that always then imagines that body as a location for transgressive pleasure. (bell hooks 1994, 131)

For this atonement, on which hinges the Christian hope of salvation, Northern Renaissance art found the painfully intimate metaphor of the Father's hand on the groin of the Son breaching a universal taboo as the fittest symbol of reconciliation. (Leo Steinberg 1996 [1983], 106)

The world spectator is emphatically a desiring subject. He has done more than accede to lack; he has learned to take pleasure in his own insatiability. (Kaja Silverman 2000, 11)

As the scholarship summarized and juxtaposed here suggests, "race" is, in the white male mind, "gender" deferred and displaced. Through its historically specific and culturally variegated metamorphosis,[2] race has, alas, assumed its own autonomous materiality, traveled along its own political and economic trajectories. By some estimates, the casualties number 100 million. Millions live the material effects today, among them poverty, educational underachievement, and an ongoing and deadly "crisis of black masculinity" (see Pinar 2001, 855 ff.) The genie cannot be stuffed back into the bottle or, more precisely, back inside the tent, but might it be pedagogically subverted within whiteness by discerning its gendered genesis? Such a curricular imperative is not performed as an "evacuation" (Wiegman 1995, 163) of race, but, rather, as a self-shattering occupation of race (their own) by European Americans.

My pedagogical strategy recalls Stuart Hall's (1996, 19) analysis of Fanon's project: to "subvert" the representational "structures" of "othering" by juxtaposing the sacred and secular in white male subjectivity, hoping that such heresies enable students to "constitute new subjectivities, new positions of enunciation and identification." Such "conscious resignification" requires, Eng (2001, 80) suggests, "unconscious support," the introduction of the "forbidden material of unconscious prohibitions." Studying these prohibitions, as we have here, may challenge those "complex networks of signifying chains"—here they are religion, race, and father–son incest—that, Eng (2001, 78) argues, have been created as unconscious in order to enable the "disguised" (deferred and displaced) "expression" of a "prohibited desire" [incest], "object" [the body of the father], or [the sodomitical primal] "scene."

The clues to this convoluted matrix lie in the gendered metaphors employed to represent racial politics and violence. "Emasculation" and "castration" communicate the character of black suffering, bell hooks (1996, 82) explains, and such suffering is the "pain of men inflicted upon them by other men." For Fanon, hooks (1996, 82) notes, "healing . . . takes place only as this conflict between men is resolved." Such healing will not occur through white men's imaginative rendering of it, as Robyn Wiegman (see 1995, 149 ff.) makes clear.

Recalling Foucault, Robyn Wiegman (1995, 116) underscores that the "disciplinary specularity" of the lynching-castration "scene" mutates, during the twentieth-century (albeit too slowly to save the lives of many other victims), to "other practices" of "surveillance" and "containment." In the twentieth century, Wiegman argues, African-American representational inclusion in popular and political culture—obvious in the struggles over representation in the school curriculum (see Zimmerman 2002)—became the primary sites in the economies of racial visibility.

What becomes visible, bell hooks (1994, 131) asserts, is the black male body as the "embodiment of bestial, violent, penis-as-weapon hypermasculine assertion." In contemporary and especially visual forms of commodification, hooks continues, this perceived threat becomes "diffused" through its fetishized representation. "[T]hrough a process of patriarchal objectification," she suggests, the black male body is rendered "feminine." She quotes Melody Davis (1991, 67; quoted in hooks 1994, 131) to emphasize her point: "specularized, men will lose their potency and force . . . they will be subject as are women to conditions, like pregnancy, beyond their control . . . they will become the sign for exchange value, and, as is the custom for women, be mere objects, voids for the gaze." Specularization spells the effacement of black subjectivity as it conflates blackness with hypersexualized embodiment, a recapitulation of the structure of servitude compelled by Noah's curse.

What if we were to redirect sexualized specularity toward white men? That is the possibility Michel Foucault, most prominently, has considered (see Bersani 1995, 77). Leo Bersani is unconvinced. Although acknowledging that S/M performs the continuity between political structures of oppression and the body's erotic economy, Bersani suspects that breaking that continuity changes neither. (For women the situation is, perhaps, different: see Cvetkovich 2003, 56.) Those male practitioners and defenders of S/M who, like Foucault, seem to believe that authoritarian structures could somehow be dissolved if only we enact the desires that support them, thereby exchanging a fraternally fascistic homosociality for self-deconstructing homosexuality. "It is as if," as Bersani (1995, 90) reconstructs the S/M argument, "recognizing the powerful appeal of those structures, their harmony with the body's most intense pleasures, they were suggesting that we substitute for history a theatricalized imitation of history." Such theater—the specular but self-reflexive performance of our scripts—allows us, presumably, to walk away from such "interpellations" into a more socially horizontal political organization.

"[I]n S/M," Bersani (1995, 90) summarizes, "we can step out of the roles whenever we like." Fantasy is, presumably, extracted from history. In contemporary racial economies of visibility, Robyn Wiegman argues, the opposite is true: history is extracted, only fantasy remains.

"Mass-mediated visual technologies"—among them, the cinema—now produce racialized representation, Wiegman (1995, 116) notes. Jonathan Crary (1999, 12) dates the appearance of "a modernizing mass visual culture . . . [to] the late 1870s." Siobhan Somerville (2000, 10) locates the genesis of this development between the 1890s and the 1920s, during which time a "number" of "new visual technologies" emerged, "particularly the development of cinema as a popular medium." As one expression of the "surveillance" of "bodies" that was "embedded" in "discourses" of expertise such as "sexology," Somerville (2000, 10) asserts, these technologies reflected and structured a "profound reorganization of vision and knowledge in American culture." The imbrication of sexuality and race posed, Somerville continues, "representational problems" concerning the "physical legibility of identity," and, as a consequence, "the emergent film industry in the United States [both] articulated and simultaneously evaded links between racial difference and homosexuality" (2000, 10). These links, and their simultaneous articulation and evasion, become visible in late-twentieth-century interracial buddy films (see Pinar 2001, 1108), cinematic representations of "miscegenation without sex" (Castronovo 2001, 244).

Although progress in the material and political circumstances of African Americans has occurred over the last century, white resistance and racism persist, and in mutated forms. The late-nineteenth-century white obsession with the black male body as sexually predatory—the nearly omnipresent image of the black male rapist—morphs, in the late twentieth century, into a white obsession with the black male body as sexually appealing: the nearly omnipresent image of the black male "stud." This "vapid fetishization" of the "visible," as Wiegman (1995, 116) so precisely puts it, is hardly limited to the sphere of the sexual.

Wiegman asserts that representational integration, in both popular culture and in the literary canon, has displaced the civic question of political power. It is a point made in a different context by Hazel Carby (1998, 191):

> These intimate black and white male partnerships, which exclude women, project the black masculinity imagined by white male liberals in quest of perfect partners. Together and alone, these race men of Hollywood dreams promise to annihilate what ails this nation and resolve our contemporary crisis of race, of nation, and of manhood.

White men's failure—refusal—to share political power in the public sphere is recoded representationally on the big screen as homosocial interracial solidarity.

Rather than sharing political power, rather than making reparations for slavery, segregation and ongoing racism, European Americans substitute a specular interracial homosociality, enabling them to persist in their commodification of the black body. The "difficult demand" of "Afrocentric political critique," Wiegman (1995, 116) observes, has been translated into "strategies" for multiplying capital's "consumer needs." Wiegman (1995, 117) notes, with understatement, that the black body's commodity status today is "not without irony" given the history of race in the United States: "the literal commodification of the body under enslavement is now stimulated in representational circuits that produce and exchange subjectivities through the visible presence of multicultural skin." Still in servitude, the racialized son remains indispensable to the white father, if banished from the tent, now relocated to the big screen. Such representation, Kaja Silverman (1988, 10) points out (in a somewhat different context), not only "covers over the absent real with a simulated or constructed reality, it also makes good the spectating subject's lack, restoring him or her to an imaginary wholeness." It is, Wiegman specifies, the white spectator who fantasies wholeness through such economies of interracial visibility.

Such "multicultural skin" depersonalizes the subjectivity it encloses, rendering it wrapped, like a homosocial (white) sheet, around numerous male bodies, fabricating what Wiegman (see 1995, 151) terms figures of interracial fraternity. She refers first to Leslie Fiedler's famous assertion, first made in 1948 and repeated in 1960, that the quintessential American myth was predicated on the "mutual love of *a white man and a colored*" (1948, 146; quoted in Wiegman 1996, 151). In *The Gender of Racial Politics and Violence in America* I devote an entire chapter to summarizing Fiedler's argument, emphasizing the slippage between "love" (Fiedler would later focus on "homoeroticism") and sex in the queer constitution of "race" in America (see Pinar 2001, chapter 18). Wiegman emphasizes Fiedler's distinction, namely that, for Fiedler, it is homeroticism's "love" sans "lust," indeed, homoeroticism's (apparent) "disembodiment" that comprises its "utopian telos" (Wiegman 1995, 151). If the interracial bond is without sex, then, why, Wiegman (see 1995, 151) asks, did Fiedler start by linking the black man to the homosexual?

In locating the U.S. literary tradition "between sentimental life in America and the archetypal image . . . in which a white and a colored American male flee from civilization into each other's arms" (1960, 12; quoted in Wiegman 1995, 152), Wiegman points out that Fiedler—despite his effort to make the distinction—implies that the interracial bond is a homosexual one, specifically that the (male) escape from the constrictions of (white) culture compels an "unambiguous movement" toward "homosexual commitment" (1995, 153, 154). Contrary to the implications of the language he employs, Fiedler has constructed a clear and irrevocable choice: innocent homoerotic bonding one on hand and "adult homosexual love" on the other (1960, 12; quoted in Wiegman 1995, 154). Wiegman (1995, 154) points out that such a developmental differentiation positions the homoerotic in an "imaginary" and "pre-symbolic realm"; it has the "force" of "originary desire" somehow "uncontaminated" by culture. Fiedler's fantasy may follow from the melancholy produced by the path not taken (see Butler 1997; Hope 1994).

The chaste love of men circumvents the threat of miscegenation (linked legally in the United States to incest: see Sollors 1997), Wiegman (1995, 155) points out, in a "representational circuit" in which the white—not black—woman symbolizes "racial threat." The absence of black women is predictable. If, in terms of masculinized racial politics, the "white woman" is an imaginary and inverted displacement of white men's sexualized desire for black men, it is "she" who must be "escaped" for white men to experience their decidedly "unmutual" (i.e. unreciprocated) "love" for the "colored." Within these "queer" racial politics, the black woman figures not at all, except as a stand-in for the black man, as the lynching of Mary Turner illustrates.[3]

Not only does the fantasy of an "innocent" interracial bond keep homosexuality in the closet, Wiegman (1995, 157) complains, it elides the misogyny implicit in the "evacuation" of the "feminine" as "precursor" to "interracial achievement." Because the "white woman" was—too often remains—a figment of the white male mind, her "evacuation" was not difficult. It was also necessary, as she stood—or should we "lay" given the white male obsession with black male rape—between men, black and white. As Wiegman herself notes, the two becomes conflated "others" in the white male mind. The black male becomes Adam's "rib" now "(re)integrated" thanks to *Brown v. Board of Education*.[4] Wiegman (1995, 160) argues that in Fiedler's work racially distinct men are relegated to the status of both "symbols" and "symptoms" of the "psychodrama" of "white masculinity." In so doing, Fiedler recodes the "historical agency" of African-American "protest" into a "sentimental male bonding relation" (1995, 160), sentimental only from a white man's point of view.

Wiegman (1995, 191) articulates what could be a key moment in the shattering of whiteness, namely the "retrieval" of a "complicated psychic interiority" that cannot be reduced those "subjective determinations" interpellated by the "social scripting" of race and gender as "corporeal visibilities." To go back inside the tent is to re-enter the disavowed body now reduced to its social surface, to its (white) social utility, to retrieve "a compli-cated psychic interiority" obviously absent in European-American mass culture. The "challenge," Wiegman (1995, 191) writes, resides in the "resignification" of the relationship between the body's interiority and its surface. As Wiegman demonstrates, it is this historical, cultural, and, I would underline, specifically "religious" relationship that underwrites the white displacement of black subjectivity onto the surface of the black body in the white eye. "Blackness" is more than "skin deep," Wiegman (1995, 192) reminds; racialization itself is "epistemically" inseparable from the articulation of other differences, especially, she notes, of gender. It is underwritten, I am suggesting, by religious repudiations of originary sexual and gendered differences within men.

It is through the law of the Father and the law of castration, as Lee Edelman and David Eng (see 2001, 128) have pointed out, that the (re)construction of the primal scene is retroactively structured as a project of heterosexual identification. The traumatizing knowledge of the primal scene for the heterosexual male viewing it in (be)hindsight is precisely the memory of its sodomitical uncertainty, that, like Schreber, he, too, may have been the apple of his Father's eye. The genealogical descendents of Shem and Japheth—"straight," especially "white" men—decline to see the naked truth. Enslaved and lynched black bodies populate the "material history of the [white] unconscious" (Castronovo 2001, 20–21).

Sexual and gender difference imagined as coinciding with anatomical differences—between, presumably, "opposite" sexes, that is, between "men" and "women"—requires, as we saw in chapter 5, "Modern Masculinity is a Stereotype," an "economy of visibility" (Wiegman 1995, 195). It is an economy, recalling the curse of Ham, that "casts social subjectivity as constitutive of the flesh" (Wiegman 1995, 195). In the binary signification of "sex" as "colored" and "race" as "sexual," social and human subjectivities were segregated within European culture. Moreover, in the disciplinary technologies associated with modernity, as these became systematized as classifications, the body's "race" and "gender" were employed, Wiegman (1995, 195) points out, as "indexes" of "psychic interiority" itself.

Specularity replaces subjectivity. The "visible" serves as the "signifying structure," Wiegman (1995, 195) notes, for the black body's apparently "evacuated interior domain." Specularity precipitates and subjectively restructures in servitude the son's evacuation from the body of the father, inside the father's tent. This paternal repudiation of the son, Wiegman (1995, 195) continues (speaking of Harriet Beecher Stowe but making my point as well), constitutes the "radical negation" compelled by "domination." The abjection of self-same desire becomes alterity, racialized, materialized in the body of the son in servitude. Subjectivity becomes invisible as alterity is visualized. The body of the father becomes the race of the son.

BACK IN THE GARDEN

I'm less persuaded . . . that queerness opens up a path "out of whiteness." (Mason Stokes 2001, 183)

It is not by willful self-naming that we shall find the exit from the prison-house of phallogocentric language. (Rosi Braidotti 1994a, 51)

The future of whiteness looks black. (Gary Taylor 2002, 140)

Robyn Wiegman (1995, 172) posits that, in the literary tradition (or counter-tradition) she has examined—not only the work of Leslie Fiedler, but that of Robert K. Martin and Joseph Boone as well—the "radical displacement" of "heterosexual romance" becomes both a "precondition to" and "symbolic enactment of" white men's fantasies of "racial transcendence." Wiegman suggests that the interracial male bond's defiance of the history of racial politics and violence in America "subverts" the "heterosexual model" of "social interaction" by converting "alienation" and "differentiation" into "mutuality" and "sameness" (1995, 172). This fantasy ignores the "complexities" of "interracial male bonds" by imagining that the escape from "heteronormativity" enables white men to also overcome racism, classism, sexism (1995, 172).[5] I agree, but as I have asked: from where, genealogically, does this fantasy derive?

In my view, the biblical fantasy of sexual difference—imagined by men as coinciding with anatomical differences (exaggerated by cultural custom into "opposite" sexes, obscuring, for instance, the omnipresence of "intersexed" individuals)—was the precursor to the European conception of racial difference. Sexual difference followed from self-severance of the self-same body, difference created by the God-the-Father to convert father–son incestuous desire into procreative sexuality. The incest taboo is the trace of what became an obsession with property, reproduction, and genealogy. Given this gendered economy, it was no accident that it was a rib ("one of the paired curved bony or partly cartilaginous rods that stiffen the walls of the body of most vertebrates and protect the viscera") and not some other organ (the eye?) that was transfigured into an(other) human form.[6] In the primal scene of racialized servitude in the white male mind—Genesis 9:23—the sexualized son (his father's rib) is cursed with anatomical, gendered, and political difference.

The pedagogical potential of racialized self-shattering such a curriculum offers is limited and uncertain. Certainly it must be severed from any utopian fantasies of interracial male solidarity, as Wiegman (see 1995, 173) cautions. This fantasy obscures the convoluted complexities of interracial bonds by glossing over the historical genealogies of racial formation. As a moment in the "talking cure" that is curriculum as cultural psychoanalysis, however, studying the desire encoded in abjection might trigger repressed memory of history. Without Wiegman's trenchant analysis, fantasy, not memory, is more likely stimulated.

Reflecting on men's interracial bond films in recent decades—*Enemy Mine*[7] is the film discussed in detail in *American Anatomies*—Wiegman (1995, 174) points out that popular representations of "inclusion" in our time transpose black demands for "political representation" into the "fetishistic display" of blackness as "commodity." In such specular conversions, Wiegman continues, the visibility of black presence in the white popular cultural representation simulates (and thereby substitutes for) historical and political subjectivities. In such specular integration, Wiegman (1995, 175) notes, the formation of a "collective" black identity, compelled by slavery and, later, inspired by the black power and civil rights movements, is "routinely exchanged" for a "hollow, historically vacated subjectivity."

This same "exchange" can be transacted in those whiteness studies courses in which attitudes and platitudes, not historical specificities, comprise the syllabus. It is the very sexual-racial narcissism and exhibitionism[8] of whiteness that appears immune to abstraction, whiteness itself being an abstraction. Not until the subjective structures that "see" racial difference only on the surface of bodies are shattered can white men glimpse the subjective materialities of others. Without subjective reconstruction, there can be no subverting of the curse; racial equality remains, for whites, a psychological not political reality (see Castronovo 2001, 201).

In a curriculum of self-shattering, white men are no longer propped up by black men, materially, culturally or psychologically. Forced to face—in the mirror—the incestuous sadomasochistic specularity of racialized alterity, European American men might make reparations. To the curriculum to which this book makes, I hope, a modest contribution, the black man no longer serves as the "enabling figure" (Wiegman 1995, 177) for the white male's traumatic rebirth, "healed" by the devoted black male friend whom he had before "misjudged, hated, or feared" (1995, 177). In that fantasy, a new white man emerged, but, as Wiegman (1995, 177) underscores, still the "voice of authority," a "trembling beacon" of "democratic ideals." Democracy remains a "purely psychological aspiration best deferred for some future day" (Castronovo 2001, 202–203).

Such an (en)raptured and racist reconfiguration depends, Wiegman has demonstrated, on continued commodification of the African-American male, enslaved still in an economy of visibility, if now structured by the specular relations of late-twentieth-century technological–cultural production. Likewise, the black male body circulates in academic discourse, specularized through its rhetorical tip of the hat toward "visibility" and "inclusion," wherein the "circulation" of the black body testifies to its "emancipatory, post-segregationist appeal" (1995, 177). Such "inclusivist" scholarly gestures—in particular Wiegman (1995, 185) decries the "monosyllabic, infinitely appended gender, race, and class"—sign that a hegemonic whiteness remains secure.

The politics of inclusion remains a politics of incorporation, in which "difference" is reconfigured not as "absolute alterity" but according to its use-value, in scholarship less economic and sexual but political still. The desire to be the subject who can "control" the politics of discourse, who can "guarantee" the consequence of one's pronouncements, who can "totally" and "finally *know*," is, Wiegman (1995, 185) points out, a "powerful desire," one that "betrays," through its demands for "mastery," the "partialities" and shifting "contingencies" that surround and saturate "cultural productions." It is a reassertion of racialized mastery.

Despite that postmodern acceptance that knowledge is not truth, that truths are never politically neutral, that subjectivity is never the transcendent ground of knowledge or truth, many whites seem unaware that what Wiegman (1995, 185) terms the "integrationist strategy"—adding black women to feminist histories is her example—expresses this racialized (white) desire for power and knowledge. Although hardly unmindful of the importance of reconstructing Western knowledges about minoritized peoples, Wiegman (see 1995, 190–191) is suspicious of the effect. She points out that categorical constructions such as "women of color" reassert, even as they resist, racist and ethnocentric logics of visibility in the West. It is the panoptical will to knowledge God-the-Father—that metaphysical substitute for white men—requires.

THE SHATTERING OF WHITENESS

Can identity itself be renegotiated in the force field where "race" and sexuality are each inflected by the other's gravitational pull? (Lee Edelman 1994b, 59)

What if the materials of memory are overwhelming, so traumatic that the remembering of them threatens identity rather than reconstituting it? (W.J.T. Mitchell 1994, 200)

Psychoanalysis challenges us to imagine a nonsuicidal disappearance of the subject. (Leo Bersani 1995, 99)

"Reality" is promiscuous. (Wahneema Lubiano 1996, 183)

The curse of the covenant will not be dispelled in one lifetime, let alone one semester. Our calling is to study it. It is to return to the genealogical recesses of European-American culture, to re-experience its archaic structures, specifically the covenant between father and son. Such "study," Alan A. Block (2004, 2) tells us, is "a way of being—it is an ethics." In contrast to Christian culture's bifurcation of knowledge and action, Block (2004, 2) argues that, within rabbinical traditions, "study . . . is a stance we assume in the world." Split off from the world, study leads to fantasy, not to history, as Robyn Wiegman makes clear.

I have no hope (see Morris 2001), only determination. There seems a certain inevitability to the "covenant" between father and son, across culture, religion, and historical moment. Is circumcision the sublimated substitute for semen transfer which expresses, in ritualized form, Noah's repudiation of the desire Sambian men performed? Among the ancient Israelites, this repudiation is strident, constructed by a curse legislated by the Leviticus laws. Among Christians, the cursed and sacrificed son is hung nearly naked upon a cross, his death enabling him to become conjoined with the Father, back inside that Chuppah now fantasized as "heaven."

"For God so loved the world, that he gave his only begotten Son" (John 3:16). This was *love*? Was the son his to sacrifice? What if the son declines to be sacrificed, declines to be hung on the cross, penetrated by the soldier's sword, nails piercing hands and feet (recall that the feet represented the penis for ancient Israelites), wearing that crown of thorns, a mocking, mutilating sign of castration? What if the son sees that the crucifixion is a meaningless gesture designed to exonerate the Father? What if the "ever-lasting" life the Father promises in that verse is only his own everlasting "life" perpetuated by "faith" and "belief"? What if the son declines the sacrifice of his life enabling identification with the father and, instead, remains (bravely, stubbornly, like the (wo)man the mother's son knows himself to be) in a desublimated position vis-à-vis the desire of the Father? What if he breaks down—shatters—the character structure the curse creates? We read one set of nineteenth-century answers to these questions in the Gospel of St. Schreber.

For Schreber, the feminized male body, abjected as the body of the Jew, internalized the curse of the covenant, enabling him to disidentify with the father and experience God's desire. In the crisis of late-nineteenth-century

European masculinity, the figure of the Jew came to represent the disavowed desire of Christian fathers and their abject sons, a desire that was transgendered and culturally transposed. Was it fortuitous that the clitoris was known at this time in Viennese slang as the *Jude* or *Jew*? (see Boyarin 1997, 211; DiPiero 2002, 139). In the crisis of late-nineteenth-century European-American masculinity, African Americans represented the desire that must be segregated, contained, castrated, its origins in Genesis 9:23 obscured. Rather than feminized, black men were hypermasculinized, the other side of the same queer coin.

Stephen Haynes (2002, 203) asks us to reimagine the curse of Ham so that the dynamics of blame are "subverted." Such subversion can be accomplished, he suggests, "only when the story is read in the context of the biblical canon and its message of redemption" (2002, 203). My curricular agenda shares Haynes' interest in subversion but is, obviously, more secular and more aggressive. For me, reparation, not "redemption," is what is ethically and erotically (see Silverman 2000, 47) required. Payments to make amends, the "work of reparation" requires the "affirmation of the ineluctability of differ-ence and deferral" (Lukacher 1986, 44). Reparation requires what Kaja Silverman (1992) characterizes as the shattering or dissolution of hegemonic white masculinity, as it is that series of subject positions that has underwritten and continues to underwrite racism. Hegemonic—racist, misogynistic—white masculinity is, for me, the horrific legacy of that mythic drunken night inside Noah's tent. Voluptuousness, not renunciation, engenders reparation.

After Kaja Silverman (1992), I theorize three interrelated elements of hegemonic white masculinity: (a) the denial of the maternal body and the gendered vulnerability (lack or castration) preoedipal symbiosis symbolized; (b) displacing self-same sexual difference onto the (M)other; and (c) thereby constructing alterity, verified epistemologically and experienced sexually through specularity. In other words, self-same sexual difference denied racializes alterity through specularity: the body of the father becomes the race of the son.

Whether Ham's transgression was sexually penetrating his father or "merely" looking at the naked body of the father, in both instances he saw his "lack," an embodied state of "castration," which the father then denied in the curse. Schreber performed his gendered and racialized lack, as he suc-cumbed to God-the-Father's desire. In Noah, lack denied displaced alterity from within the self-same body, projected it onto an "other" specularized as "difference" doomed to servitude. In Schreber, alterity introjected shattered his subjective structures, rendering him unable to re-enter the world as, in Fanon's (1968, 316) utopic phrase, a "new man."

We want neither Noah nor Schreber; like Fanon, we want a "new man." The curriculum I have sketched here will not midwife the birth of a "new man," but it asks students to encounter their own alterity, specularity, and lack. It invites students to re-experience what Freud characterized as the "negative" oedipal complex, enabling a restructuring of internal object relations in which binaries are mixed and merged in the self-same (now simultaneously the "opposite" sexed) body. "When identification is non-identical," Regina Schwartz (1997, 117) has observed, "there is no motive to replace." There is no genealogical impulse, no compulsion to replicate oneself in future generations, indeed, no future in Lee Edelman's sense, in which the present is sacrificed for what never will be.

Like Silverman (1988, 154), I am arguing for the "shattering" of the white male subject, in which his narcissism and exhibitionism are exposed, thereby threatening the collapse of the ancient patriarchal scopic regime upon which sexual and racial difference relies (see also Silverman 1988, 162). In doing so, I am declining that masculine appropriation of the feminine Gerald Izenberg describes. Nor am I fantasizing a return to an originary "pure, pre-cultural," and "pre-conscious desire" (Wiegman 1995, 154) that enables, presumably, interracial bonds to be forged in a garden where, in the absence of women and civilization, racial politics disappear into nonsexual love.

Robyn Wiegman (1995, 126) is surely right to point out that the fascination with interracial male bonds as simultaneously a "democratic achievement" and an expression of the "mythic national unconscious" (1995, 172) reiterates the invisibility of "woman" and "race" in the very constitution of the national imaginary, that of "America" itself. Such class-blind, binary, womanless and raceless fantasies structure the social surface of the Western white male mind. It is, I am suggesting, by a curricular encounter with these structures that students might undertake the dissolution of whiteness.

Such an undertaking hardly promises (although teachers may be tempted by the ambition) that white students can "deconstruct their own whiteness and decolonize their Eurocentrism in order to abolish or transcend their racial significance" (Cohen 1997, 245). Whiteness and, in particular, hegemonic masculinity are too pervasive, too complicated, too unconscious for white men to be so confident. Nor will the racialized self-understanding of whiteness to which I hope this volume contributes be the "if only" the technology of education always promises (and always fails to deliver). There can be no predictable "outcomes" of serious study; there can be no science of education. What is possible is study.

Challenging the hegemony of ocularcentrism in Western (white) culture does not threaten blindness. Helpful here is art historian Norman Bryson's (1983, 94) distinction between the "gaze"—that "prolonged, contemplative, yet regarding the field of vision with a certain aloofness and disengagement, across a tranquil interval"—and that of the "glance," "a furtive or sideways look whose attention is always elsewhere, which shifts to conceal its own existence and which is capable of carrying unofficial, *sub rosa* messages of hostility, collusion, rebellion, and lust." (The gay cruise is, clearly, a subspecies of the "glance."[9]) Bryson (1983, 95) warns that when the "Gaze [is] victorious over the Glance, vision [is] disembodied [and] decarnalised." The gaze stereotypes (see Bryson 1983, 156, 159) and commodifies. Understood as the glance, "to look . . . *is* to care (Silverman 2000, 73).

Let us all look, then, at ourselves, as the palimpsest we personify. Like W.J.T. Mitchell (1994, 417), I have assembled this textbook as one might fashion a "photograph album," inspired by what Kaja Silverman (2000, 62) describes as an "ethics of desire—an ethics grounded in a passion for symbolization, in a delight in the manifold and ever new forms that the past can assume." This synoptic text is a "collection" of "snapshots" of (glances at) whiteness, a textbook addressed, especially, to white men who wish to study the stereotypical in themselves. As Mitchell says of his own, if the "book has a unity, it has been in its insistence on staying for the many answers to its few questions" (1994, 417) of whiteness. After Mitchell, I ask, what if we thought of whiteness, itself a form of representation,

> not as a homogeneous field or grid of relationships governed by a single principle, but as a multi-dimensional and heterogeneous terrain, a collage or patchwork quilt assembled over time out of fragments. Suppose further that this quilt was torn, folded, wrinkled, covered with accidental stains, traces of the bodies it has enfolded. (Mitchell 1994, 419)

These stains are not, of course, accidental: they are traces of enslaved and mutilated black bodies.

Such a model of whiteness might make materially visible the genesis of racism, whiteness as deferred and displaced self-same desire. Stripped from its originary setting, whiteness becomes intelligible as an "ongoing process of assemblage, of stitching in and tearing out," mutating into a "multi-dimensional and heterogeneous terrain," disguised even as interracial homosocial friendship. Still following Mitchell (see 1994, 410), I ask what if we thought about whiteness, not as a noun but as a verb, structuring a set of relationships? "Suppose," he continues, "we de-reified the *thing* that seems to 'stand' before us, 'standing for' something else," and thought of whiteness "as a process in which the thing is a participant, like a pawn on a chessboard or a coin in a system of exchange?" Like this expansive and dynamic notion of representation, such a conceptual move would construe whiteness as "roughly commensurate with the totality of cultural activity," including

> that aspect of political culture which is structured around the transfer, displacement, or alienation of power—from "the people" to "the sovereign," the state, or the representative, from God to father to son in a particular system, from slave to master in an absolutist polity. (Mitchell 1994, 410)

Such "cultural activity" is the sea in which we white men swim, taken for a ride on an ark by an odd old man who, after the waters have receded, is about to plant a vineyard. This time we will not accept our servitude, this time we will articulate our "language of desire" (Silverman 2000, 67). When we speak this language, Silverman (2000, 67) tells us, we come to "understand" that the "past is not yet fully written" thereby releasing us from the "paralysis of being" into the "mobility of becoming." So released, will we become other than what our Father cursed us to be? Will we make reparations? Let us see what study engenders.

NOTES

PREFACE

1. Martin Jay (1993, 508 n. 56) notes that for Derrida that light and dark constitute "the founding metaphor of Western philosophy as metaphysics. The founding metaphor not only because it is a photological one—and in this respect the entire history of our philosophy is photology, the name give to a history of, or treatise on, light—but because it is a metaphor. Metaphor in general, the passage from one existent to another, authorized by the initial submission of Being to the existent, the analogical displacement of Being, is the essential weight which anchors discourse in metaphysics" (p. 27).
2. Of course, the history of "race" is considerably more complex than that paragraph implies. During the last quarter of the eighteenth century, for instance, race was reconceptualized. Before this transitional period, Dror Wahrman (see 2004, 127) tells us, race had been basically mutable, described as changeable through the effects of climate and the environment, or, accordingly to beliefs specific to the eighteenth century, through "human interventions" in the forms of "social customs": or even "individual choice." From the 1770s onward, however, race was "gradually" if "haltingly reconceptualized" as an "essential and immutable category, stamped on the individual" (Wahrman 2004, 127). Moreover, the invocation of the Curse of Ham declined during late eighteenth century in England (see Wahrman 2004, 199), but not so in the United Sates (see Haynes 2002).
3. I will not argue that the restructuring of white masculinity requires a regression to infantile symbiosis. After all, as Kaja Silverman (1988, 160) points out, "distance from the mother is the precondition not only of subjectivity and language, but of desire itself." "Until about 8 months," Joseph H. Smith (2004, 350) acknowledges, "I am my mother. The shock of not being one's mother is the context in which the question emerges of who, then, am I." The repudiation of this preoedipal identification with the mother displaces desire (see Silverman 2000, 122) which then circulates as homosociality. Homosociality props up whiteness, as Mason Stokes (2001, 18) appreciates: "[T]he homosocial may be a necessary component of any attempt to keep whiteness white, to keep whiteness pure." It is homosexuality, Scott Derrick (1997, 223 n. 27) understands, that can "disrupt the narcissism of male homosocial mirroring."
4. The psychoanalytic notion of "deferred action" (*Nachtraglichkeit*) is a term Freud employed to explain how the experience of trauma is deferred—and, I would add, displaced—into other subjective and social spheres, including racial spheres, where it is no longer readily recognizable. It is the precursor, I suspect, to Lacan's "most important contribution to psychoanalysis," in Margarita Palacios' (2004, 292) judgement, namely his "conceptualization of the

unconscious and the metonymic character of desire." Contrary to the transparent subjectivity of the Cartesian cogito, to the dialogical subjectivity of symbolic interactionism, and to the normalizing subjectivity of functionalism and post-structuralism, Palacious (2004) points out, Lacanian subjectivity includes within itself its own "impossibility," a concept structuring Lee Edelman's critique of the mythic—and specifically political—status of the "child" in contemporary U.S. politics (2005). Faced with the failure of symbolization, Palacios (2004, 292) tells us, Lacan defines the unconscious as a "radical other." In the present context, that "radical other" is the racialized son.

5. Bret Hinsch (see 1992, 52) points out that prior to the Han dynasty (206 B.C.E. to 220 C.E.), a Chinese man could legally kill his own son, and traces of this life-and-death authority of the father over the son remained during the Han dynasty. There is, of course, a long tradition of male "homosexuality" in China; it is known as the passion of "cut sleeve" (see Hinsch 1990).

6. Certainly, the hegemony of ocularcentrism was challenged during the twentieth century, especially in France (Jay 1993). Jonathan Crary (1999, 3) acknowledges: "At the present moment, to assert the centrality or 'hegemony' of vision within twentieth-century modernity no longer has much value or significance at all."

7. God is, of course, the "Big Other," the "biggest man of all" (Driver 1996, 51).

Introduction

1. Discussing Luce Irigaray ("whose argument carries great force"), Silverman (1988, 185) explains that men's negative Oedipus complex has primacy, "at least at the level of the unconscious, over its positive counterpart." In fact, the positive Oedipus complex can be understood as the "indirect" and "disguised expression" of the son's unacknowledged (and unacknowledgeable) desire for the father, the exchange of women finally a "pretext for putting man in touch with man." Irigaray's argument suggests, Silverman (1988, 185) concludes, that the "phallus is what the penis becomes when it itself cannot be enjoyed."

2. Crary (1999, 3) does not believe that "exclusively visual concepts such as the 'gaze' or 'beholding' are in themselves valuable objects of historical explanation."

1   In the Beginning

1. Sander Gilman (2005, B15) notes "Abrahamic" is the "new buzzword" that includes Islam in the "Judeo–Christian fold."

2   Inside the Tent

1. Richard Dyer (1997, 28) suggests that "the divided nature of white masculinity . . . reproduces the structure of feeling of the Christ story." Christ's agony was that he was "fully flesh" and "fully spirit," tempted by sin but able to resist. In the "torment" of the crucifixion, Dyer continues, Christ experienced the "fullness of the pain of sin," transcended in the resurrection. In the scene of the crucifixion, Dyer (1997, 28) asserts, "the spectacle of white male bodily suffering typically conveys a sense of the dignity and transcendence

in such pain." For David Savran, this sadomasochistic structure of masculinity enables us to "take it like a man" (1998).

2. And not only racialized, but nationalized: the two conflate, as George Mosse (1985, 41) has observed. Rey Chow (see 2002, 24) notes that the Greek word *ethnos* means nation or people.

3. Leo Steinberg (1996) studied early Renaissance representations of Christ's sexuality, specifically, artistic representations of his penis:

> If the motif of the self-touching infant seems rash enough, even more at odds with what one expects of devotional art is the Christ Child's erection. . . . The earliest Italian instance of the motif known to me is a *Madonna and Child with Four Angels* by Giovanni dal Ponte (Florentine, c. 1385–1437) in the De Young Museum, San Francisco . . . . (1996 [1983], 183)

The association between the nakedness of Noah and of the Christ child becomes obvious when Steinberg (1996 [1983], 43 n. 33) cites Martin Schongauer's use of the robe parted motif to expose Jesus' genitals, commenting that:

> The motif of the robe parted to expose the sex may have its Christian inspiration in the traditional image of the *Drunkenness of Noah*, itself conceived as a type of Passion. In the *Presentation* altarpiece by the late 15th-century German Master of the Life of the Virgin (National Gallery, London), the Noah scene on the depicted retable over the altar appears as a prefiguration behind the nude Christ Child.

3 THE SPECULARITY OF ALTERITY

1. Crary (see 1999, 3) argues that ideas about vision cannot be separated from a larger historical restructuring of subjectivity that involved not only optical experiences but processes of modernization and rationalization.

2. Wiegman is well acquainted with the main texts, as her bibliographic list (see 1995, 3) attests.

4 OUTSIDE THE TENT

1. This is, of course, a well-known fact. I reference here Marla Morris' important study of teaching the Holocaust because she grapples so brilliantly with both curriculum and cultural issues, interwoven as they are. Luther becomes intelligible, for instance, as both descendent and progenitor of anti-Semitic Christian culture.

5 DECADENCE, DISORIENTATION, DEGENERATION

1. "The beginnings of sexology," Somerville (2000, 31) points out, "circulated within and perhaps depended on a pervasive climate of eugenicist and anti-miscegenation sentiment and legislation."

2. Also writing about the same period, Scott Derrick (1997, 157) argues that:

> The intense desire for the redeeming stuff of masculinity is shadowed by a growing panic that experientially, masculine affiliations, enthusiastically

cultivated, cannot be separable in kind from erotic desires for other men. Such a threat is in part contained by the casting of the homosexual as a recognizable figure antithetical to normative masculinity, but this defense against homoeroticism paradoxically also intensifies the threat of impermissible, feminizing desires alien to the specular, manifest masculinity of the touchdown run, the heroic charge, or the Western duel.

3. Illustrative of Mosse's point is Richard Dyer's (1997, 170) observation that "Italian fascism's imaginary of masculinity centered on monumentalist imagery," producing "massive statuary and painting (especially frescoes and posters) featuring big men in aspirant postures." Surrounding the Mussolini-built sports stadium in Rome, the Foro Italico, is, Dyer (1997, 170) points out, "statues three times life-size of muscular naked men in white stone."

4. Fanon (1967) argued the oedipal problem did not exist for Africans, including Africans in the diaspora; others have argued it may not exist for Chinese: see Wang 2004.

6   An Epistemology of the Body

1. Hinsch (see 1992, 147) notes that among surviving examples of Chinese erotic art are representations that, curiously, "distinguish active and passive partners through tint—with the skins of the active partner taking on a darker tone."

7   The Curse of the Covenant

1. Helga Geyer-Ryan (1996, 122) posits the nation—after the house and the city—as the "body-ego's third double," underscoring the "prelogical" and "prelinguistic" character of xenophobia. The presence of strangers, Geyer-Ryan (1996, 122) continues, can be experienced as an "assault" on and a "violent penetration" of the "unified" but "always fragile" and "precarious body-self," an event that threatens the "fragmentation" and "collapse" of the "imaginary autonomous ego." The pedagogical point is for white men to experience this "collapse" as the self-shattering dissolution of whiteness.

2. The democratization of intra-subjective relations, as Kobena Mercer (1994, 232) acknowledges, requires "a new black queer cultural politics . . . the democratizing of our desires in all their diversity and perversity."

3. Likewise, Richard Dyer (1997, 78) asserts: "Whiteness, really white whiteness, is unattainable."

4. "The gentilizing [gentile] and whitening of Christ," Dyer (1997, 68) notes, "was achieved by the end of the Renaissance and by the nineteenth century the image of him as not just fair-skinned but blond and blue-eyed was fully in place."

8   The Sodomitical Subjectivity of Race

1. "Voluptuousness undoes all schemas," Cathryn Vasseleu (1996, 132) asserts, "all thematization of the world; it is a beginning without memory, a beginning that knows no other . . . . It has no basis in the subject that sees things but is a state of immersion."

2. Now that "race" is blocked in the political discourse of the American South, white men revert to gender. So-called conservative values—Black and Black (1992, 9) list "traditional family values, the importance of religion, support for capital punishment, and opposition to gun control" and I would add opposition to abortion and opposition to civil rights for lesbians and gay men— mobilize a defensive white masculinity and, thereby, preserve traces of earlier racist recalcitrance. As such, they constitute "deferred and displaced" versions of racism. The reactionary energy that animates white Southerners' sometimes fanatical (on occasion, even homicidal) engagement with these "values" reveals the presence of the racist past (see Pinar 2004a, chapter 4). Although Southern "conservatism" cannot be reduced to residues of racial hatred—it is broader and more complex than that—it cannot be understood apart from it either.

3. The events leading to Mary Turner's lynching started with a white farmer in south Georgia who refused to pay his black employee the wages due to him. A few days later the unscrupulous farmer was found shot to death. Unable to find the man who had motive for the murder, white mobs began to kill every African American who had even the remotest connection with the victim and the alleged slayer. One of those murdered by a white mob was a black man named Hayes Turner, whose crime was that he knew the accused; both men had worked for the dead farmer. Turner's wife, Mary, was grief-stricken; she cried out in sorrow, cursing those who had left her a widow and her unborn child fatherless. She threatened to swear out warrants to bring her husband's murderers to justice (White, 1929).

Her husband's murderers learned of her threat. "We'll teach the damn nigger wench some sense," they responded, and began to search for her. Understanding her peril, her friends hid the grieving woman on a obscure farm, miles away. It was on a Sunday morning, "with a hot May sun beating down," Walter White (1929) reports, when they found her. White (1929), who went to investigate the crime, describes the scene:

> Securely they bound her ankles together and, by them, hanged her to a tree. Gasoline and motor oil were thrown upon her dangling clothes; a match wrapped her in sudden flames. Mocking, ribald laughter from her tormentors answered the helpless women's screams of pain and terror. The clothes burned from her crisply toasted body, in which, unfortunately, life still lingered. A man stepped towards the woman and, with his knife, ripped open the abdomen in a crude Cesarean operation. Out tumbled the prematurely born child. Two feeble cries it gave—and received for answer the heel of a stalwart man, as life was ground out of the tiny form. Under the tree of death was scooped a shallow hole. The rope about Mary Turner's charred ankles was cut, and swiftly her body tumbled into its grave. Not without a sense of humor or of appropriateness was some member of the mob, as an empty whisky-bottle, quart size, was given for headstone. Into its neck was stuck a half-smoked cigar—which had saved the delicate nostrils of one member of the mob from the stench of burning human flesh. (White 1929, 28–29; see also Hernton 1988 [1965], 129; Pinar 2001, 91–92)

4. See Wiegman's (1995, 157 ff.) placement of Fielder's thesis within the decade of school integration. She suggests that Fielder has internalized within Anglo male subjectivity the race war that was Little Rock.

5. As I have complained about "whiteness" studies in education (Pinar 2001), the specificity of historical experience disappears in presentistic abstractions (e.g., "teaching" for "tolerance" or against "prejudice"). Psychoanalytically oriented curriculum theorists have underscored the complexity and intensity of resistance to difficult knowledge, resistance subjectively sedimented and socially enacted, and including in classrooms (Pitt 2003; Britzman 1998a, b, 2003). Mitchell (1994, 201) suggests:

> The refusal to "go back" in memory, triggered by the request to recall a color, is a refusal to revive a visual memory, to remember the experience in a form that brings it too close, too near to a re-experiencing of the unspeakable.

What could that "unspeakable" be?

6. It is tempting to play with Webster's definition, of course, as the social roles of "woman" within patriarchy have so often been structured to protect (and elicit) the "soft" side of men, enabling their public personae to be "stiff."

7. This film (dir. Wolfgang Petersen), Wiegman (1995, 126) points out, portrays the "dark buddy" as "quite literally" an alien. A reptilian figure, Jerry (played by Louis Gossett Jr.) stands in stark contrast to Willis Davidge (Dennis Quaid), the alien's "altogether human" (1996, 126) white male companion. In casting the black man as alien and his white "bonding buddy" as human, Wiegman (1996, 126) continues, the movie "conjures up" a long and outrageous tradition of black stereotypes, among them the dark beast; the "corporeal essence is defined and symbolized according to the logic of the visible, and hierarchical arrangements are naturalized in the dyadic relationship between identity and difference."

8. Through an "extraordinary sleight of hand," Kaja Silverman (1988, 26) notes, women has been made the "repository" not only of "lack" but of "specularity." (As have black men and women, I might remind.) Silverman adds that women have also, in this same sleight of hand, come to be identified with "narcissism" and "exhibitionism," qualities, she points out, "more compatible" with male subjectivity than with female, qualities "almost synonymous" with "organ display." Certainly, in the lynching ritual the black man's organ was on display; as Wiegman's analysis requires us to notice, the black body remains on display today, especially cinematically.

9. Likewise, Bryson's distinction between gaze and glance, Van Alphen (1996, 172) points out, enables us to analyze the "particular modes of homosocial looking as powerful forms of social agency and, by extension, as configurations of the homosocial cultural order as such."

References

Alexander, Elizabeth (1994). "Can you be BLACK and look at this": Reading the Rodney King Video(s). In Thelma Golden (Ed.), *Black Male: Representations of Masculinity in Contemporary American Art* (91–110). New York: Whitney Museum of American Art (Harry N. Abrams, Inc.).

Armstrong, Karen (2001). *The battle for god: A history of fundamentalism.* New York: Ballantine.

Aron, L. (1995). The internalized primal scene. *Psychoanalytic Dialogues* 5 (2), 195–237.

Baker, Bernadette M. (2001). *In perpetual motion: Theories of power, educational history, and the child.* New York: Peter Lang.

Barth, Fredrik (1987). *Cosmologies in the making: A generative approach to cultural variation in Inner New Guinea.* Cambridge: Cambridge University Press.

Bederman, Gail (1995). *Manliness and civilization: A cultural history of gender and race in the United States, 1880–1917.* Chicago: University of Chicago Press.

Bell, Vikki (1993). *Interrogating incest: Feminism, Foucault and the law.* London: Routledge.

Benjamin, Jessica (1995). *Like subjects, love objects: Essays on recognition and difference.* New Haven, CT: Yale University Press.

Benjamin, Jessica (1998). *In the shadow of the other.* New York: Routledge.

Berger, John (1972). *Ways of seeing.* New York: Viking.

Bersani, Leo (1994). Is the rectum a grave? In Jonathan Goldberg (Ed.), *Reclaiming Sodom* (249–264). New York: Routledge.

Bersani, Leo (1995). *Homos.* Cambridge, MA: Harvard University Press.

Bettelheim, Bruno (1955). *Symbolische wunden* [*Symbolic wounds.*] Franfurt am Main: Frischer Taschenbuch Verlag.

Black, Earl and Black, Merle (1992). *The vital South: How presidents are elected.* Cambridge, MA: Harvard University Press.

Block, Alan A. (2001). Ethics and curriculum. *JCT* 17 (3), 23–38.

Block, Alan A. (2004). *Talmud, curriculum, and the practical: Joseph Schwab and the Rabbis.* New York: Peter Lang.

Boone, Joseph A. (1987). *Tradition counter tradition: Love and the form of fiction.* Chicago: University of Chicago Press.

Bordo, Susan (1993). *Unbearable weight: Feminism, western culture, and the body.* Berkeley: University of California Press.

Bordo, Susan (1994). Reading the male body. In Laurence Goldstein (Ed.), *The Male Body* (265–306). Ann Arbor: University of Michigan Press.

Bowie, Malcolm (1991). *Lacan.* Cambridge, MA: Harvard University Press.

Boyarin, Daniel (1997). *Unheroic conduct: The rise of heterosexuality and the invention of the Jewish man.* Berkeley: University of California Press.

Braidotti, Rosi [with Butler, Judith] (1994a). Interview: Feminism by an other name. *Differences* 6 (2+3), 27–61.

Braidotti, Rosi (1994b). Revisiting male thanatica. *Differences* 6 (2+3), 199–207.

Brakke, David (2001, October). Ethiopian demons: Male sexuality, the black-skinned other, and the monastic self. *Journal of the History of Sexuality* 10 (3–4), 501–535.

Britzman, Deborah P. (1998a). *Lost subjects, contested objects: Toward a psychoanalytic inquiry of learning.* Albany, NY: State University of New York Press.

Britzman, Deborah P. (1998b). Is there a queer pedagogy? Or, stop reading straight. In William F. Pinar (Ed.), *Curriculum: Toward New Identities* (211–231). New York: Garland.

Britzman, Deborah P. (2003). *After-education: Anna Freud, Melanie Klein, and psychoanalytic histories of learning.* Albany: State University of New York Press.

Brown, Roger (1965). *Social psychology.* New York: The Free Press.

Brownmiller, Susan (1993 [1975]). *Against our will: Men, women, and rape.* New York: Fawcett Columbine.

Bruno, Giuliana (1994). The body of Pasolini's semiotics: A sequel twenty years later. In Patrick Rumble and Bart Testa (Eds.), *Pier Paolo Pasolini: Contemporary perspectives* (88–105). Toronto: University of Toronto Press.

Bryant, Levi R. (2004). Politics of the virtual. *Psychoanalysis, Culture & Society* 9, 333–348.

Bryson, Norman (1983). *Vision and painting: The logic of the gaze.* New Haven, CT: Yale University Press.

Burton, Robert V. and Whiting, J.M.W. (1961). The absent father and cross-sex identity. *Merrill-Palmer Quartery* 7, 85–95.

Buruma, Ian (1984). *Behind the mask: On sexual demons, sacred mothers, transvestites, gangsters, drifters and other Japanese cultural heroes.* New York: Pantheon.

Butler, Judith (1997). *The psychic life of power: Theories in Subjection.* Stanford, CA: Stanford University Press.

Bynum, Caroline Walker (1982). *Jesus as mother.* Berkeley: University of California Press.

Campbell, J.K. (1964). *Honor, family, and patronage.* Oxford: Clarendon.

Canetti, Elias (1978). *Crowds and power.* [Trans. Carol Stewart.] New York: Seabury Press.

Carby, Hazel V. (1998). *Race men.* Cambridge, MA: Harvard University Press.

Carnes, Mark C. (1990). Middle-class men and the solace of fraternal ritual. In Mark C. Carnes and Clyde Griffen (Eds.), *Meanings for Manhood: Constructions of Masculinity in Victorian America* (37–52). Chicago: University of Chicago Press.

Castenell, Jr., Louis A., and Pinar, William F. (1993). Introduction. In Louis A. Castenell, Jr. and William F. Pinar (Eds.), *Understanding curriculum as racial text: Representations of identity and difference in education* (1–30). Albany, NY: State University of New York Press.

Castronovo, Russ (2001). *Necro citizenship: Death, eroticism, and the public sphere in the nineteenth-century United States.* Durham, NC: Duke University Press.

Cavanagh, Sheila L. (2004). Upsetting desires in the classroom: School sex scandals and the pedagogy of the femme fatale. *Psychoanalysis, Culture & Society* 9, 315–332.

Chodorow, Nancy J. (1978). *The reproduction of mothering.* Berkeley: University of California Press.

Chow, Rey (2002). *The Protestant ethnic & the spirit of capitalism.* New York: Columbia University Press.

Cleaver, Eldridge (1968). *Soul on ice.* New York: Dell Publishing Co.

Cohen, H. Hirsch (1974). *The drunkenness of Noah*. University, AL: University of Alabama Press.

Cohen, Phil (1997). Laboring under whiteness. In Ruth Frankenberg (Ed.), *Displacing Whiteness: Essays in Social and Cultural Criticism* (242–282). Durham, NC: Duke University Press.

Comstock, Gary David (1991). *Violence against lesbians and gay men*. New York: Columbia University Press.

Crary, Jonathan (1990). *Techniques of the observer: On vision and modernity in the nineteenth century*. Cambridge, MA: MIT Press.

Crary, Jonathan (1999). *Suspensions of perception: Attention, spectacle, and modern culture*. Cambridge, MA: The MIT Press.

Creed, Gerald W. (1994). Sexual subordination: Institutionalized homosexuality and social control in Melanesia. In Jonathan Goldberg (Ed.), *Reclaiming Sodom* (66–94). New York: Routledge.

Cvetkovich, Ann (2003). *An archive of feelings: Trauma, sexuality, and lesbian public cultures*. Durham, NC: Duke University Press.

Daly, Mary (1978). *Gyn/ecology: The metaethics of radical feminism*. Boston: Beacon.

Dance, L. Janelle (2001). *Tough fronts*. New York: RoutledgeFalmer.

Davis, Melody D. (1991). *The male nude in contemporary photography*. Philadelphia: Temple University Press.

Deacon, Bernard A. (1934). *Malekula: A vanishing people in the New Hebrides*. London: G. Routledge & Sons, Ltd.

de Beauvoir, Simone (1974 [1952]). *The second sex*. [Trans. H.M. Parshley.] New York: Vintage.

de Bolla, Peter (1996). The visibility of visuality. In Teresa Brennan and Martin Jay (Eds.), *Vision in Context: Historical and Contemporary Perspectives on Sight* (63–81). New York: Routledge.

de Lauretis, Teresa (1994). Habit changes. *Differences* 6 (2+3), 296–313.

Derrick, Scott S. (1997). *Monumental anxieties: Homoerotic desire and feminine influence in nineteenth-century U.S. literature*. New Brunswick, N.J.: Rutgers University Press.

Dewey, John (1991 [1927]). *The public and its problems*. Athens: Ohio University Press.

Dewey, John (1960). *On experience, nature and freedom*. (Edited by Richard Bernstein.) Indianapolis: Library of Liberal Arts.

DiPiero, Thomas (2002). *White men aren't*. Durham, NC: Duke University Press.

Doane, Mary Ann (1982). Film and the masquerade—theorizing the female spectator. *Screen* 23 (3–4), 74–88.

Doane, Mary Ann (1991). Veiling over desire: Close-ups of the woman. In Richard Feldstein and Judith Roof (Eds.), *Feminism and Psychoanalysis*. Ithaca, NY: Cornell University Press.

Dollimore, Jonathan (1991). *Sexual dissidence: Augustine to Wilde, Freud to Foucault*. Oxford: Oxford University Press.

Douglass, Frederick (1972 [1853]). The heroic slave. In Ronald T. Takaki (Ed.), *Violence in the Black Imagination: Essays and Documents* (37–77). New York: G.P. Putnam's.

Dover, Kenneth J. (1978). *Greek homosexuality*. Cambridge, MA: Harvard University Press.

Driver, Tom F. (1996). Growing up Christian and male. In Bjorn Krondorfer (Ed.), *Men's bodies, men's gods: Male identities in a (post-) Christian culture* (43–64). New York: New York University Press.

Dyer, Richard (1997). *White*. London: Routledge.

Edelman, Lee (1994a). Seeing things: Representation, the scene of surveillance and the spectacle of gay male sex. In Jonathan Goldberg (Ed.), *Reclaiming Sodom* (265–287). New York: Routledge.

Edelman, Lee (1994b). *Homographesis: Essays in gay literary and cultural history*. New York: Routledge.

Edelman, Lee (2004). *No future: Queer theory and the death drive*. Durham, NC: Duke University Press.

Eilberg-Schwartz, Howard (1994). *God's phallus. And other problems for men and monotheism*. Boston: Beacon Press.

Ellison, Ralph (1995 [1964/1972]). *Shadow and act*. New York: Random House. [Reprinted in 1972/1995 by Vintage.]

Eng, David L. (2001). *Racial castration: Managing masculinity in Asian America*. Durham, NC: Duke University Press.

Fanon, Frantz (1967). *Black skin, white masks*. [Trans. by Charles Lam Markmann.] New York: Grove Weidenfeld. [Originally published in French under the title *Peau Noire, Masques Blancs*, copyright 1952 by Editions du Seuil, Paris.]

Fanon, Frantz (1968). *The wretched of the earth*. [Preface by Jean-Paul Sartre. Trans. by Constance Farrington.] New York: Grove Press. [Originally published by François Maspero éditeur, Paris, France, under the title *Les damnés de la terre*, 1961.]

Fichte, Hubert (1996). *The gay critic*. [Trans. by Kevin Gavin. Introduction by James W. Jones.] Ann Arbor: University of Michigan Press.

Fiedler, Leslie A. (1948). "Come back to the raft ag'in, Huck Honey!" *Partisan Review* 15, 664–711.

Fiedler, Leslie A. (1966 [1960]). *Love and death in the American novel*. [Revised edition.] New York: Stein & Day.

Flynn, Thomas R. (1993). Foucault and the eclipse of vision. In David Michael Levin (Ed.), *Modernity and the Hegemony of Vision* (273–286). Berkeley: University of California Press.

Foucault, Michel (1973 [1966]). *The order of things*. New York: Vintage.

Foucault, Michel (1995 [1975]). *Discipline and punish*. [Trans. Alan Sheridan.] New York: Pantheon.

Foucault, Michel (1976). *History of sexuality*. [Trans. Robert Hurley.] New York: Vintage.

Foucault, Michel (1990 [1988]). *Politics, philosophy, culture: Interviews and other writings 1977–1984*. [Edited with an introduction by Lawrence D. Kritzman.] New York and London: Routledge.

Freud, Sigmund (1911). Psychoanalytic notes on an autobiographical account of a case of Paranoia. *Standard Edition of the Complete Psychological Works* (12:59). [Trans. James Strachey. 24 vols.] London: Hogarth.

Freud, Sigmund (1912–1913). Totem and taboo. *Standard Edition of the Complete Psychological Works* (13:1–161). [Trans. James Strachey. 24 vols.] London: Hogarth.

Freud, Sigmund ([1913] 1946). *Totem and taboo*. [Trans. A.A. Brill] New York: Vintage.

Freud, Sigmund (1918 [1914]). From the history of an infantile neurosis. *Standard Edition of the Complete Psychological Works* (17:7–122). [Trans. James Strachey. 24 vols.] London: Hogarth.

Freud, Sigmund (1923 [1922]). A seventeenth-century demonological neurosis. *Standard Edition of the Complete Psychological Works* (19:69–108). [Trans. James Strachey. 24 vols.] London: Hogarth.

Freud, Sigmund (1927). The ego and the id. *Standard Edition of the Complete Psychological Works* (21:3–58). [Trans. James Strachey. 24 vols.] London: Hogarth.

Freud, Sigmund (1933). *New introductory lectures on psychoanalysis.* London: Hogarth.

Freud, Sigmund (1939). *Moses and monotheism.* London: Hogarth Press.

Freud, Sigmund (1950). *Totem and taboo.* London: Routledge and Kegan Paul.

Freud, Sigmund (1953–1974). *The standard edition of the complete psychological works.* [Trans. James Strachey. 24 vols.] London: Hogarth Press.

Freud, Sigmund (1957 [1914]). On Narcissism. *Standard Edition of the Complete Psychological Works* (73–102). [Trans. James Strachey. vol. 14] London: Hogarth.

Freud, Sigmund (1960). *The ego and the id.* [Trans. Joan Riviere.] New York: Norton.

Freud, Sigmund (1963). *Three case histories.* New York: Collier.

Freud, Sigmund (1968 [1953]). Totem and taboo. Volume 13. *Standard Edition of the Complete Psychological Works.* [Trans. James Strachey. 24 vols.] London: Hogarth.

Gadamer, Hans-Georg (1975). *Truth and method.* (Trans. Garrett Barden and John Cumming.) New York: Seabury Press.

Gardiner, Muriel (Ed.) (1971). *The Wolf-Man by the Wolf-Man.* New York: Basic Books.

Gartner, Richard B. (1999). *Betrayed as boys: Psychodynamic treatment of sexually abused men.* New York: Guilford.

Gates, Jr., Henry Louis (1988). *The signifying monkey: A theory of Afro-American literary criticism.* New York: Oxford University Press.

Gay, Peter (1988). *Freud: A life for our times.* New York: Norton.

Gay, Peter (1993). *The cultivation of hatred.* New York: Norton.

Geller, Jay (1993). Freud v. Freud: Freud's reading of Daniel Paul Schreber's *Denkwürdigkeiten eines Nervenkranken.* In Sander Gilman, Juta Birmele, Valerie Greenberg, and Jay Geller (Eds.), *Reading Freud's reading.* New York: New York University Press.

Geyer-Ryan, Helga (1996). Imaginary identity: Space, gender, nation. In Teresa Brennan and Martin Jay (Eds.), *Vision in Context: Historical and Contemporary Perspectives on Sight* (118–125). New York: Routledge.

Gilman, Sander L. (1993). *Freud, race, and gender.* Princeton, NJ: Princeton University Press.

Gilman, Sander L. (2005, April 8). The parallels of Islam and Judaism in diaspora. *The Chronicle* LI (31), B15–B16.

Gilmore, David (1990). *Manhood in the making.* New Haven: Yale University Press.

Gilmore, David (2001). *Misogyny: The male malady.* Philadelphia: University of Pennsylvania Press.

Gilroy, Paul (1993). *The black Atlantic.* Cambridge, MA: Harvard University Press.

Goldhill, Simon (1996). Refracting classical vision: Changing cultures of viewing. In Teresa Brennan and Martin Jay (Eds.), *Vision in Context: Historical and Contemporary Perspectives on Sight* (15–28). New York: Routledge.

Gollaher, David L. (2000). *Circumcision: A history of the world's most controversial surgery.* New York: Basic Books.

Gordon, Lewis R. (1995). *Bad faith and antiblack racism.* Atlantic Highlands, NJ: Humanities Press.

Gordon, Lewis R. (1996). Can men worship? Reflections on male bodies in bad faith and a theology of authenticity. In Bjorn Krondorfer (Ed.), *Men's bodies, men's gods: Male identities in a (post-) Christian culture* (235–250). New York: New York University Press.

Gorsline, Robin Hawley (1996). Facing the body on the cross: A gay man's reflection on passion and crucifixion. In Bjorn Krondorfer (Ed.), *Men's bodies, men's gods: Male identities in a (post-) Christian culture* (125–145). New York: New York University Press.

Goux, Jean-Joseph (1992). The phallus: Masculine identity and the "exchange of women." [Trans. by Maria Amuchastegui, Caroline Benforado, Amy Hendrix, and Eleanor Kaufman.] *Differences* 4 (1), 40–75.

Goux, Jean-Joseph (1993). *Oedipus, philosopher.* [Trans. Catherine Porter.] Stanford, CA: Stanford University Press.

Graves, Robert and Patai, Raphael (1964). *Hebrew myths: The book of Genesis.* Garden City, NY: Doubleday.

Greene, Naomi (1990). *Pier Paolo Pasolini: Cinema as heresy.* Princeton, NJ: Princeton University Press.

Greven, Philip (1977). *The Protestant temperament: Patterns of child-rearing, religious experience, and the self in early America.* Chicago: University of Chicago Press.

Gregor, Thomas (1985). *Anxious pleasures: The sexual life of an Amazonian people.* Chicago: University of Chicago Press.

Griswold, Robert L. (1998). The "flabby American," the body, and the cold war. In Laura McCall and Donald Yacovone (Eds.), *A Shared Experience: Men, Women, and the History of Gender* (323–348). New York: New York University Press.

Grossberg, Michael (1990). Institutionalizing masculinity: The law as a masculine profession. In Mark C. Carnes & Clyde Griffen (Eds.), *Meanings for Manhood: Constructions of Masculinity in Victorian America* (133–151). Chicago: University of Chicago Press.

Grosz, Elizabeth (1994). The labors of love. Analyzing perverse desire: An interrogation of Teresa de Lauretis's *The Practice of Love. Differences* 6 (2+3), 274–295.

Grosz, Elizabeth (1995). *Space, time, and perversion: Essays on the politics of bodies.* New York: Routledge.

Grumet, Madeleine R. (1988). *Bitter milk: Women and teaching.* Amherst: University of Massachusetts Press.

Hall, Stuart (1996). The after-life of Frantz Fanon: Why Fanon? Why now? Why black skin, white masks? In Alan Read (Ed.), *The fact of blackness: Frantz Fanon and visual representation* (12–37). Seattle: Bay Press.

Halperin, David M. (1990). *One hundred years of homosexuality.* New York: Routledge.

Hannaford, Ivan (1996). *Race: The history of an idea in the West.* [Foreword by Bernard Crick.] Baltimore, MD: Johns Hopkins University Press.

Harris, Cheryl I. (1993). Whiteness as property. *Harvard Law Review* 106 (8), 1707–1791.

Hartman, Saidiya V. (1997). *Scenes of subjection: Terror, slavery, and self-making in nineteenth century America.* New York: Oxford University Press.

Haynes, Carolyn A. (1998). *Divine destiny: Gender and race in nineteenth-century Protestantism.* Jackson: University Press of Mississippi.

Haynes, Stephen R. (2002). *Noah's curse: The biblical justification of American slavery.* New York: Oxford University Press.

Herdt, Gilbert H. (1981). *Guardians of the flute.* New York: McGraw-Hill.

Herdt, Gilbert H. (1982). Fetish and fantasy in Sambia initiation. In Gilbert H. Herdt (Ed.), *Rituals of Manhood* (44–98). Berkeley: University of California Press.

Hernton, Calvin C. (1988 [1965]). *Sex and racism in America.* New York: Doubleday (Anchor).

Herzfeld, Michael (1985). *The poetics of manhood: Contest and identity in a Cretan mountain village.* Princeton: Princeton University Press.

Hinsch, Bret (1990). *Passions of the cut sleeve: The male homosexual tradition in China.* Berkeley: University of California Press.

Hoch, Paul (1979). *White hero, black beast: Racism, sexism and the mask of masculinity.* London: Pluto Press.

Hocquenghem, Guy (1978). *Homosexual desire.* London: Allison and Busby.

Hodgkin, T. (1957). *Nationalism in colonial Africa.* New York: New York University Press.

Hoffman, Lawrence A. (1996). *Covenant of blood: Circumcision and gender in Rabbinic Judaism.* Chicago: University of Chicago Press.

Hofstadter, Richard (1962). *Anti-intellectualism in American life.* New York: Vintage.

hooks, bell (1994). Feminism inside: Toward a black body politic. In Thelma Golden (Ed.), *Black Male: Representations of Masculinity in Contemporary American Art* (127–140). New York: Whitney Museum of American Art (Harry N. Abrams, Inc.).

hooks, bell (1996). Feminism as a persistent critique of history: What's love got to do with it? In Alan Read (Ed.), *The fact of blackness: Frantz Fanon and visual representation* (76–85). Seattle: Bay Press.

Hope, Trevor (1994). Melancholic modernity: The hom(m)osexual symptom and the homosocial corpse. *Differences* 6 (2+3), 174–198.

Houlgate, Stephen (1993). Vision, reflection, and openness. The "hegemony of vision" from a Hegelian point of view. In David Michael Levin (Ed.), *Modernity and the Hegemony of Vision* (87–123). Berkeley: University of California Press.

Huebner, Dwayne E. (1999). *The lure of the transcendent.* Mahweh, NJ: Lawrence Erlbaum.

Hunt, Alan (1998). The great masturbation panic and the discourses of moral regulation in nineteenth- and early twentieth-century Britain. *Journal of the History of Sexuality* 8 (4), 575–615.

Izenberg, Gerald N. (2000). *Modernism and masculinity: Mann, Wedekind, Kandinsky through World War I.* Chicago: University of Chicago Press.

Jay, Martin (1988a). *Fin-de-siecle socialism and other essays* (52–63). New York: Routledge.

Jay, Martin (1988b). Scopic regimes of modernity. In Hal Foster (Ed.), *Vision and Visuality* (3–23). Seattle, WA: Bay Press.

Jay, Martin (1993). *Downcast eyes: The denigration of vision in twentieth-century French thought.* Berkeley: University of California Press.

Jordan, Neil (Dir.) (1994). *Interview with the vampire.* Starring Tom Cruise and Brad Pitt. Based on the novel by Ann Rice. Warner Studios, 123 minutes.

Judovitz, Dalia (1993). Vision, representation, and technology in Descartes. In David Michael Levin (Ed.), *Modernity and the Hegemony of Vision* (63–86). Berkeley: University of California Press.

Katz, Jonathan Ned (1994). The age of sodomitical sin, 1607–1740. In Jonathan Goldberg (Ed.), *Reclaiming Sodom* (43–58). New York: Routledge.

Keesing, Roger M. (1982). Introduction. In Gilbert H. Herdt (Ed.), *Rituals of Manhood* (1–43). Berkeley: University of California Press.

Kelly, R.C. (1976). Witchcraft and sexual relations: An exploration in the social and semantic implications of the structure of belief. In P. Brown and G. Buchbiner (Eds.), *Man and Woman in the New Guinea Highlands.* Special publication of the American Anthropological Association.

Kelly, R.C. (1977 [1974]). *Etoro social structure: A study in structural contradiction.* Ann Arbor: University of Michigan Press.

Kimmel, Michael S. (1996). *Manhood in America: A cultural history.* New York: Free Press.

Laplanche, J. and Pontalis, J.B. (1973). *The language of psychoanalysis.* [Trans. by Donald Nicholson-Smith.] New York: Norton.

Krondorfer, Bjorn (1996). Introduction to *Men's bodies, men's gods: Male identities in a (post-) Christian culture* (3–26). New York: New York University Press.

Landtman, Gunnar (1927). *The Kiwi Papuans of British New Guinea: A nature-born instance of Rousseau's ideal community.* London: Macmillan.

Langeveld, M. (1983). The stillness of the secret place. *Phenomenology + Pedagogy* 1 (1), 11–17.

Laplanche, J. and Pontalis, J.B. (1973). *The language of psychoanalysis.* [Trans. by Donald Nicholson-Smith.] New York: Norton.

Laqueur, Thomas (1990). *Making sex: Body and gender from the Greeks to Freud.* Cambridge, MA: Harvard University Press.

Laqueur, Thomas W. (2003). *Solitary sex: A cultural history of masturbation.* New York: Zone Books.

Lasch, Christopher (1978). *The culture of narcissism: American life in an age of diminishing expectations.* New York: Norton.

Lasch, Christopher (1984). *The minimal self: Psychic survival in troubled times.* New York: Norton.

Layard, John (1942). *Stone men of Malekula.* London: Chatto & Windus.

Le Rider, Jacques (1993). *Modernity and crises of identity: Culture and society in fin-de-siecle Vienna.* [Trans. Rosemary Morris.] New York: Continuum.

Levin, David Michael (1993a). Introduction. *Modernity and the hegemony of vision* (1–29). Berkeley: University of California Press.

Levin, David Michael (1993b). Decline and fall: Ocularcentrism in Heidegger's reading of the history of metaphysics. In David Michael Levin (Ed.), *Modernity and the Hegemony of Vision* (186–217). Berkeley: University of California Press.

Loizos, Peter (1975). *The Greek gift.* New York: St. Martin's Press.

Lubiano, Wahneema (1996). "But compared to what?" Reading realism, representation, and essentialism in *School Daze, Do the Right Thing,* and the Spike Lee Discourse. In Marcellus Blount and George P. Cunningham (Eds.), *Representing Black Man* (173–204). New York: Routledge.

Lukacher, Ned (1986). *Primal scenes: Literature, philosophy, psychoanalysis.* Ithaca, NY: Cornell University Press.

Mahler, Margaret et al. (1975). *The psychological birth of the infant.* New York: Basic Books.

Mailer, Norman (1968). *Armies of the night.* New York: New American Library.

Marriott, David (2000). *On black men.* New York: Columbia University Press.

Marshall, Bill (1997). *Guy Hocqueghem: Beyond gay identity.* Durham, NC: Duke University Press.

Martin, Robert K. (1986). *Hero, captain, and stranger: Male friendship, social critique, and literary form in the novels of Herman Melville.* Chapel Hill: University of North Carolina Press.

Masters, R.E.L. (1963). *Patterns of incest: A psycho-social study of incest based on clinical and historic data.* New York: The Julian Press.

McCumber, John (1993). Derrida and the closure of vision. In David Michael Levin (Ed.), *Modernity and the Hegemony of Vision* (234–251). Berkeley: University of California Press.

McLaren, Andrew (1997). *The trials of masculinity: Policing sexual boundaries, 1870–1930*. Chicago: University of Chicago Press.

Meggitt, Mervyn J. (1977). *Blood is their argument*. Palo Alto: Mayfield Pub. Co.

Mercer, Kobena (1994). *Welcome to the jungle: New positions in black cultural studies*. New York: Routledge.

Mercer, Kobena and Julian, Issac (1988). Race, sexual politics, and black masculinity: A dossier. In Rowena Chapman and Jonathan Rutherford (Eds.), *Male Order: Unwrapping Masculinity* (97–164). London: Lawrence and Wishart.

Mirsky, Seth (1996). Three arguments for the elimination of masculinity. In Bjorn Krondorfer (Ed.), *Men's bodies, men's gods: Male identities in a (post-) Christian culture* (27–39). New York: New York University Press.

Mitchell, W.J.T. (1994). *Picture theory: Essays on verbal and visual representation*. Chicago: University of Chicago Press.

Moglen, Helene (1997). Redeeming history: Toni Morrison's *Beloved*. In Elizabeth Abel, Barbara Christian, and Helene Moglen (Eds.), *Female Subjects in Black and White* (201–220). Berkeley: University of California Press.

Morris, Marla (2001). *Holocaust and curriculum*. Mahwah, NJ: Lawrence Erlbaum.

Morris, Marla and Weaver, John (Eds.) (2002). *Difficult memories: Talk in a (post) Holocaust Era*. New York: Peter Lang.

Mosse, George L. (1985). *Nationalism and sexuality: Respectability and abnormal sexuality in modern Europe*. New York: Howard Fertig.

Mosse, George L. (1996). *The image of man: The creation of modern masculinity*. New York: Oxford University Press.

Mottram, R.H. (1937). *Noah*. London: Rich & Cowan, Ltd.

Mudimbe, V.Y. (1988). *The invention of Africa: Gnosis, philosophy, and the order of knowledge*. Bloomington: Indiana University Press.

Mulvey, Laura (1985). *Visual and Other Pleasures*. Bloomington: Indiana University Press.

Munro, Petra (1998). *Subject to fiction*. Buckingham and Philadelphia: Open University Press.

Murray, Stephen O. and Roscoe, Will (1998). *Boy-wives and female husbands*. New York: Palgrave, St. Martin's Press.

Musil, Robert (1955 [1906]). *Young Torless*. [Preface by Alan Pryce-Jones.] New York: Pantheon Books Inc.

Musil, Robert (1979 [1960–1961]). *The man without qualities*. [Foreword by Elithne Wilkins. Trans. Ernst Kaiser.] London: Secker & Warburg.

Nelson, James B. (1996). Epilogue. In Bjorn Krondorfer (Ed.), *Men's bodies, men's gods: Male identities in a (post-) Christian culture* (311–318). New York: New York University Press.

Newman, Philip and Boyd, David J. (1982). The making of men: Ritual and meaning in Awa male initiation. In Gilbert H. Herdt (Ed.), *Rituals of Manhood* (239–285). Berkeley: University of California Press.

Newman, Saul (2004). Interrogating the master: Lacan and radical politics. *Psychoanalysis, Culture & Society* 9, 298, 314.

Nietzsche, Friedrich (1967). *The will to power*. [Trans. Walter Kaufman.] New York: Random House.

Nordau, Max (1895 [1892]). *Degeneration*. New York: Appleton. [Reissued in 1993 by the University of Nebraska Press.]

Nunberg, Herman (1949). *Problems of bisexuality as reflected in circumcision*. London: Imago Publishing Co., Ltd.

Oakeshott, Michael (1959). *The voice of poetry in the conversation of mankind.* London: Bowes & Bowes.

Obholzer, Karin (1982). *The Wolf-Man: Conversations with Freud's patient—sixty years later.* [Trans. Michael Shaw.] New York: Continuum.

Palacios, Margarita (2004). On sacredness and transgression: Understanding social antagonism. *Psychoanalysis, Culture & Society* 9, 284–297.

Park, Robert (1950). *Race and culture: Essays in the sociology of contemporary man.* Glencoe: Free Press.

Pasolini, Pier Paolo (1968). *The ragazzi.* [Trans. by Emile Capouya.] New York: Grove Press, Inc. [Originally published in 1955, by Aldo Garzanti Editore, Milan, Italy.]

Pasolini, Pier Paolo (1982). *Poems.* [Selected and translated by Norman MacAfee. Foreward by Enzo Siciliano.] New York: Random House.

Pasolini, Pier Paolo (1985). *A violent life.* [Trans. by William Weaver.] Manchester, UK: Carcanet Press Ltd. [First published in U.K. in 1968; published in Italy in 1959 by Aldo Garzanti Editore.]

Pfeil, Fred (1995). *White guys.* London: Verso.

Pinar, William F. (2001). *The gender of racial politics and violence in America: Lynching, prison rape, and the crisis of masculinity.* New York: Peter Lang.

Pinar, William F. (2002a). The medicated body: Drugs and Dasein. In Sherry Shapiro and Svi Shapiro (Eds.), *Body Movements: Pedagogy, Politics, and Social Change* (283–315). Cresskill, NJ: Hampton Press.

Pinar, William F. (2002b). Robert Musil and the crisis of European culture in America. In William E. Doll, Jr. and Noel Gough (Eds.), *Curriculum Visions* (102–110). New York: Peter Lang.

Pinar, William F. (2002c). Foreword. Marla Morris and John Weaver (Eds.), *Difficult Memories: Talk in a (Post) Holocaust Era* (xi–xvi). New York: Peter Lang.

Pinar, William F. (2003). Inside Noah's tent: The sodomitical genesis of "race" in the Christian imagination. In Peter Pericles Trifonas (Ed.), *Pedagogies of Difference* (155–187). New York: RoutledgeFalmer.

Pinar, William F. (2004a). *What is curriculum theory?* Mahwah, NJ: Lawrence Erlbaum.

Pinar, William F. (2004b). The problem of the public. In Rubén A. Gastambide-Fernandez and James T. Sears (Eds.), *Curriculum Work as a Public Moral Enterprise* (119–126). Lanham, MD: Rowman & Littlefield.

Pinar, William F. (2004c). Rocket man. *JCT* 20 (2), 7–13.

Pinar, William F. (2004d). The synoptic text today. *JCT* 20 (1), 7–22.

Pinar, William F. (2004e). Curriculum and study. Paper presented to the Curriculum and Pedagogy Conference, Miami University of Ohio, October.

Pinar, William F. (2005a). From Statesmanship to status: The Absence of Authority in Contemporary Curriculum Studies. *Journal of the American Association for the Advancement of Curriculum Studies* 1 (1). Available at <www.uwstout.edu/soe/jaaacs> p. 36.

Pinar, William F. (2005b). The queer character of lynching. Paper presented to the Nineteenth Century Studies Association, Baton Rouge, April.

Pinar, William F. (2006). *The synoptic text today and other essays: Curriculum development after the Reconceptualization.* New York: Peter Lang.

Pinar, William F. and Irwin, Rita L. (Eds.) (2004). *Curriculum in a new key: The collected works of Ted Aoki.* Mahwah, NJ: Lawrence Erlbaum.

Pinar, William F., Reynolds, William M. Slattery, Patrick, and Taubman, Peter M. (1995). *Understanding curriculum.* New York: Peter Lang.

Pine, Richard (1995). *The thief of reason: Oscar Wilde and modern Ireland*. New York: St. Martin's Press.

Pitt, Alice J. (2003). *The play of the personal*. New York: Peter Lang.

Pronger, Brian (1990). *The arena of masculinity: Sports, homosexuality and the meaning of sex*. New York: St. Martin's Press.

Poole, Fitz John P. (1982). The ritual forging of identity: Aspects of person and self in Bimin-Kuskusmin male initiation. In Gilbert H. Herdt (Ed.), *Rituals of Manhood* (99–154). Berkeley: University of California Press.

Ravitch, Diane (2000). *Left back: A century of battles over school reform*. New York: Simon and Schuster.

Read, Kenneth (1965). *The high valley*. London: Allen and Unwin.

Roediger, David R. (1994 [1991]). *Wages of whiteness: Race and the making of the working class*. London: Verso.

Rogin, Michael Paul (1985). "The sword became a flashing vision": D.W. Griffith's *The Birth of A Nation. Representations* 9, 150–195.

Rogoff, Irit (1996). "Other's others": Spectatorship and difference. In Teresa Brennan and Martin Jay (Eds.), *Vision in Context: Historical and Contemporary Perspectives on Sight* (187–202). New York: Routledge.

Ronell, Avital (1992). *Crack wars*. Lincoln: University of Nebraska Press.

Rosen, Jeffrey (2000). *The unwanted gaze: The destruction of privacy in America*. New York: Random House.

Santner, Eric L. (1996). *My own private Germany: Daniel Paul Schreber's secret history of modernity*. Princeton, NJ: Princeton University Press.

Sass, Louis A. (1992). *Madness and modernism: Insanity in the light of modern art, literature, and thought*. New York: Basic Books.

Sass, Louis A. (1994). *The paradoxes of delusion: Wittgenstein, Schreber, and the schizophrenic mind*. Ithaca, NY: Cornell University Press.

Savran, David (1998). *Taking it a like a man: White masculinity, masochism, and contemporary American culture*. Princeton, NJ: Princeton University Press.

Scarce, Michael (1997). *Male on male rape*. New York: Insight Books/Plenum Publishing Corporation.

Schatzman, M. (1973). *Soul murder: Persecution in the family*. New York: Random House.

Schieffelin, E. (1976). *The sorrow of the lonely and the burning of the dancers*. New York: St. Martin's Press.

Schieffelin, Edward L. (1982). The *bau a* ceremonial hunting lodge: An alternative to initiation. In Gilbert H. Herdt (Ed.), *Rituals of Manhood* (155–200). Berkeley: University of California Press.

Schreber, Daniel Paul (2000 [1903]). *Memoirs of my nervous illness*. [Introduction by Rosemary Dinnage. Trans. and edited by Id Macalpine and Richard A. Hunter.] New York: New York Review Books. [An earlier version of this same translation—first published in 1955 but with different pagination—was published by Harvard University Press in 1968 and again in 1988; Eilberg-Schwartz, Santner and Sass quote from this English-language edition. Santner translates certain passages himself from the German 1985 edition; see Santner 1996, 147, n. 1.]

Schwartz, Regina M. (1997). *The curse of Cain: The violent legacy of monotheism*. Chicago: University of Chicago Press.

Sedgwick, Eve Kosofsky (1985). *Between men: English literature and male homosocial desire*. New York: Columbia University Press.

Sedgwick, Eve Kosofsky (1990). *Epistemology of the closet*. Berkeley: University of California Press.

Selden, Steven (1999). *Inheriting shame: The story of eugenics and racism in America.* New York: Teachers College Press.

Seshadri-Crooks, Kalpana (1998). The comedy of domination: Psychoanalysis and the conceit of whiteness. In Christopher Lane (Ed.), *The Psychoanalysis of Race* (353–379). New York: Columbia University Press.

Silverman, Kaja (1988). *The acoustic mirror: The female voice in psychoanalysis and cinema.* Bloomington: Indiana University Press.

Silverman, Kaja (1992). *Male subjectivity at the margins.* New York: Routledge.

Silverman, Kaja (2000). *World Spectators.* Stanford, CA: Stanford University Press.

Simpson, Mark (1994). *Male impersonators: Men performing masculinity.* [Foreword by Alan Sinfield.] New York: Routledge.

Smith, Joseph H. (2004). Prejudices as instances of the negative. *Psychoanalysis, Culture & Society* 9, 349–346.

Sollors, Werner (1997). *Neither black nor white yet both: Thematic explorations of interracial literature.* New York: Oxford University Press.

Somerville, Siobhan B. (2000). *Queering the color line: Race and the invention of homosexuality in American culture.* Durham, NC: Duke University Press.

Spillers, Hortense J. (1987). Mama's baby, papa's maybe: An American grammar book. *Diacritics* 17 (2), 65–81.

Stack, Oswald (1969). *Pasolini on Pasolini: Interviews with Oswald Stack.* Bloomington, IN: Indiana University Press.

Steinberg, Leo (1996 [1983]). *The sexuality of Christ in Renaissance art and in modern oblivion.* [Second edition.] Chicago: University of Chicago Press.

Stokes, Mason (2001). *The color of sex: Whiteness, heterosexuality, & the fictions of white supremacy.* Durham, NC: Duke University Press.

Stoller, Robert (1974). Facts and fancies: An examination of Freud's concept of bisexuality. In Jean Strouse (ed.) *Women and Analysis* (343–364). New York: Dell.

Stoltenberg, John (1989/1990). *Refusing to be a man.* Portland, OR: Breitenbush Books. [Reprinted in 1990 by Meridian, New York.]

Taylor, Gary (2002). *Castration: An abbreviated history of western manhood.* New York: Routledge.

Testa, Bart (1994). To film a gospel . . . and advent of the theoretical stranger. In Patrick Rumble and Bart Testa (Eds.), *Pier Paolo Pasolini: Contemporary Perspectives* (180–209). Toronto: University of Toronto Press.

Theweleit, Klaus (1987). *Male fantasies.* [Trans. S. Conway, Erica Carter, and Chris Turner.] Minneapolis: University of Minnesota Press.

Toews, J.E. (1997). Refashioning the masculine subject in early modernism: Narratives of self-dissolution and self-construction in psychoanalysis and literature, 1900–1914. *MODERNISM/modernity* 4 (1), 31–67.

Tolson, Andrew (1977). *The limits of masculinity: Male identity and the liberated woman.* New York: Harper and Row.

Trueit, Donna (2005). *Complexifying the poetic: Toward a poeisis of curriculum.* Baton Rouge: Louisiana State University, unpublished Ph.D. dissertation.

Tuzin, Donald F. (1982). Ritual violence among the Ilahita Arapesh: The dynamics of moral and religious certainty. In Gilbert H. Herdt (Ed.), *Rituals of Manhood* (321–355). Berkeley: University of California Press.

Tyack, David and Hansot, Elizabeth (1990). *Learning together: A history of coeducation in American schools.* New Haven, CT: Yale University Press.

Twitchell, James B. (1987). *Forbidden partners: The incest taboo in modern culture.* New York: Columbia University Press.

Van Alphen, Ernst (1996). The homosocial gaze according to Ian McEwan's *The Comfort of Strangers*. In Teresa Brennan and Martin Jay (Eds.), *Vision in Context: Historical and Contemporary Perspectives on Sight* (169–185). New York: Routledge.

Van Baal, J. (1966). *Dema*. The Hague, Netherlands: Martinus Nijhoff.

Vanggaard, Thorkil (1972). *Phallos*. New York: International Universities Press.

Vasseleu, Cathryn (1996). Illuminating passion: Irigaray's transfiguration of night. In Teresa Brennan and Martin Jay (Eds.), *Vision in Context: Historical and Contemporary Perspectives on Sight* (128–137). New York: Routledge.

Veyne, Paul (1985). Homosexuality in ancient Rome. In Philippe Aries and André Bejin (Eds.), *Western Sexuality* (26–39). [Trans. Anthony Forster.] Oxford: Basil Blackwell.

Wagner, R. (1972). *Habu: The innovation of meaning in Daribi religion*. Chicago: University of Chicago Press.

Wahrman, Dror (2004). *The making of the modern self: Identity and culture in eighteenth-century England*. New Haven, CT: Yale University Press.

Wang, Hongyu (2004). *The call from the stranger on a journey home: Curriculum in a third space*. New York: Peter Lang.

Ware, Vron (1992). *Beyond the pale: White women, racism and history*. London: Verso.

Warnke, Georgia (1993). Ocularcentrism and social criticism. In David Michael Levin (Ed.), *Modernity and the Hegemony of Vision* (287–308). Berkeley: University of California Press.

Weiermair, Peter (1988). *The hidden image: photographs of the male image in the nineteenth and twentieth century*. Cambridge, MA: MIT Press.

Weininger, Otto (1903/1906). *Sex and character*. London: Heinemann.

Wiegman, Robyn (1993, January). The anatomy of lynching. *Journal of the History of Sexuality* 3 (3), 445–467.

Wiegman, Robyn (1995). *American anatomies: Theorizing race and gender*. Durham, NC: Duke University Press.

West, Cornel (1993). *Race matters*. Boston: Beacon Press.

Westbury, Ian, Hopmann, Stefan, and Riquarts, Kurt (Eds.) (2000). *Teaching as reflective practice: The German didaktik tradition*. Mahweh, New Jersey: Lawrence Erlbaum Associates, Publishers.

White, Walter (1929). *Rope and faggot: A biography of Judge Lynch*. New York: Alfred A. Knopf.

Williams, F.E. (1936). *Papuans of the Trans-Fly*. Oxford: Oxford University Press.

Williamson, Joel (1984). *The crucible of race: Black-white relations in the American South since Emancipation*. New York: Oxford University Press.

Wittig, Monique (1992). *The straight mind and other essays*. Boston: Beacon Press.

Wyatt-Brown, Bertram (1982). *Southern honor: Ethics and behavior in the old South*. New York: Oxford University Press.

Yack, Bernard (1986). *The longing for total revolution: Philosophic sources of social discontent from Rousseau to Marx and Nietzsche*. Princeton, NJ: Princeton University Press.

Young, Lola (1996). *Race, gender and sexuality in the cinema*. London: Routledge.

Young-Bruehl, Elisabeth (1996). *The anatomy of prejudices*. Cambridge, MA: Harvard University Press.

Zizek, Slavoj (1998). Love thy neighbor? No, thanks!? In Christopher Lane (Ed.), *The Psychoanalysis of Race* (154–175). New York: Columbia University Press.

INDEX